HOW SCIENCE ENRICHES THEOLOGY

HOW SCIENCE
ENRICHES THEOLOGY

Benedict M. Ashley, O.P.
and John Deely

ST. AUGUSTINE'S PRESS
South Bend, Indiana

Manufactured in the United States of America

1 2 3 4 5 6   17 16 15 14 13 12

Library of Congress Cataloging in Publication Data
Ashley, Benedict M.
How science enriches theology / Benedict M. Ashley and John Deely.
p. cm.
Includes bibliographical references (p.) and index.
ISBN 978-1-58731-364-6 (clothbound: alk. paper)
1. Religion and science. 2. Catholic Church – Doctrines.
I. Deely, John N. II. Title.
BX1795.S35A84 2011
261.5'5 – dc23   2011028423

∞ The paper used in this publication meets the minimum requirements of the American National Standard for Information Sciences – Permanence of Paper for Printed Materials, ANSI Z39.48-1984.

ST. AUGUSTINE'S PRESS
www.staugustine.net

The authors wish to express special thanks for help in proofreading to Dr. Christopher Morrissey of the Philosophy Department at Redeemer Pacific College in Langley, British Columbia, Canada; the attention to detail of Professor Morrissey's corrections was so great that we can almost make him responsible for any remaining errors! Mr. Stephen Sparks of the graduate philosophy program at the University of St. Thomas, Houston, also provided help with the volume. And finally we wish to thank Professor James Clarage of the Physics Department of the University of St Thomas, Houston, for his reading of and suggestions for Chapter 2, as mentioned in note 1 of that chapter, p. 35 below.

This work is dedicated to
the Members of the Former Albertus Magnus Lyceum
and the Present Members of the Institute for Advanced Physics

Benedict M. Ashley, O.P., a graduate of the University of Chicago (where he studied with Mortimer J. Adler and Robert Hutchins) and Notre Dame University, was ordained a Dominican priest in 1948. From 1963–1969 he was the Regent of Studies at the Aquinas Institute of Philosophy at River Forest, Illinois (a Pontifical Faculty accredited also by the North Central Association), which emerged as one of the three leading centers of Thomistic thought in North America, the "River Forest School", with an emphasis on the continuity of modern science with Latin Age natural philosophy. Author of more than nineteen books and numerous articles, his *Way toward Wisdom* (University of Notre Dame Press, 2006) in particular provides an overview of the Neothomistic development launched by Pope Leo XIII in 1869. Fr. Ashley served as a Consultant for the National Catholic Bishops Conference, 1965–2000. He has been a consultant and teacher at many institutions, including the John Paul II Cultural Center in Washington, DC. Retired as Emeritus Profess of the Aquinas Institute of Theology at St. Louis University, he is currently living at St. Pius V Dominican Priory in Chicago, and is an Associate Faculty of Philosophy for the Institute for Advanced Physics in Baton Rouge, Louisiana.

John Deely holds the Rudman Chair in Philosophy at the Center for Thomistic Studies at the University of St. Thomas in Houston, Texas. Author of twenty-three books and several hundred articles, he has also co-authored and edited some thirty-four volumes, including notably *Shaping Psychology. How we got where we're going* with Msgr. Timothy Gannon (= Sources in Semiotics vol. X, 1991), and *The Problem of Evolution* (Appleton-Century-Crofts, 1969) with Fr. Raymond J. Nogar, O.P., himself author of *The Wisdom of Evolution* and *Lord of the Absurd*. Named the first living Thomas A. Sebeok Fellow of the Semiotic Society of America in 1993, in October of 2009 Deely received both the "Aquinas Medal for Excellence in Christian Philosophy" from the International Gilson Society, and the "Scholarly Excellence Award of the American Maritain Association". In 2001 he published *Four Ages of Understanding. The first postmodern survey of philosophy from ancient times to the turn of the 21ˢᵗ century* (University of Toronto Press), and most recently *Medieval Philosophy Redefined* (University of Scranton Press, 2010), showing how Latin thought provided the background and framework which made science in the modern sense an inevitable development.

CONTENTS OVERVIEW

Preface

"TRUTH CANNOT CONTRADICT TRUTH"

Reason, our intellect or "human understanding", Thomas Aquinas points out in many places,[1] is ordered to know first of all the surrounding world "outside" of our minds. The maturation of that ordering is precisely what we have come to call "science", *both* science in the sense of philosophy, which deals with the critical control of objectification within the sphere of experience common to all human beings, *and* science in the modern sense which goes beyond common experience to critically control objectification by means of experiments and mathematization of results.

As we will examine below,[2] *"scientia"* — science — in medieval usage, as earlier in Aristotle, was still a generic term, yet one anticipating in particular the specific development of modern science, "ideoscopic" science (which Aquinas considered generally dialectical or probable research, but which turned out to have considerably more demonstrative possibilities than anticipated), in contrast to philosophy as "cenoscopic" science (which Aquinas considered demonstrative or certain, supplying principles for further research, and also, as Peirce would much later show,[3] as semiosis, the very framework within which ideoscopy actually works).

---

[1] E.g., Aquinas 1266: *Summa theologiae* I, q. 87, art. 3, *et alibi passim.*

[2] See Chapter 1, p. 6f., esp. text at note 11.

[3] See Peirce 1902: *Minute Logic*, CP 1.241–242; 1908: "On the Classification of Signs", 8.342, *et alibi passim*. Cf. Deely 2008: *The Crossroad of Signs*

Thus the development of science in the modern sense is precisely a maturation of human understanding — one, we think, that has not been sufficiently recognized in ecclesiastical decrees heretofore;[4] yet the technologies to which science has given rise have become essential to advancing the well-being of our species as a whole. Truth in this sense, involving the expansion over the centuries of our understanding of the world around us, is essential to the development and well-being of civilization, as should be apparent from our dependency upon our surroundings to survive.

Now truth as it involves religious belief is something else again, for this truth does not come simply from reason, as does scientific truth, but from faith. Yet since truth is one, and does not admit of internal contradictions, the truths of faith go beyond, but cannot go contrary to, truths demonstrably attained by reason.[5] Indeed, it was the view of Aquinas that the very

---

and Ideas, p. 74 and *passim*. In the medieval context, a synonym for *scientia* in the as-yet-unadumbrated *specific* sense of cenoscopic in its contrast with specifically ideoscopic science, was precisely *doctrina* (not at all to be confused with "dogma" in the theological sense): see, in Deely 1982: Appendix I, "On the Notion 'Doctrine of Signs'," 127–130; and Deely 1986: 214, the terminological entry "Doctrine" in the *Encyclopedic Dictionary of Semiotics*, ed. Thomas A. Sebeok et al.

⁴ See, in Chapter 2, pp. 41–42 note 10, the remark of Maritain 1973: 209–210, on the error of holding ideoscopic science "in its own development to be subject to theology, and to a literal interpretation of Scripture".

⁵ Aquinas c.1257/8: *Super Boetium De Trinitate*, q. 2. art. 3c: "So therefore in the development of theological doctrine we can make use of philosophy [i.e., of *scientia*, science, generally] in three ways. First, for the purpose of demonstrating matters which are conducive to faith ['*praeambula fidei*'], which one believing ought to have knowledge of, such as the truths about God provable by natural reason — that God exists, that God is one — and other truths of this kind, whether about God or about creatures, which are proved in philosophy [or now in science in the modern sense], and which faith presupposes. Second, for calling attention to natural phenomena which suggest or bear some resemblance to truths which are held on faith, the way Augustine in his book on the Trinity draws many parallels from teachings of philosophy to illustrate the Trinity. Third, for resisting assertions advanced contrary to faith, either by showing them to be false or by showing them to be gratuitous."

development of reason in the pursuit of truth tends to suggest as true, or "points in the direction of", the very ideas that faith properly embraces as truths revealed by God to supplement and elevate and encourage the exercise of reason in the pursuit of truth and spirituality.

As we will see, St. Thomas goes so far as to explain that anyone who says that something is a sin because it offends God speaks the truth only in part, for God has so created human beings that it is impossible for a person to offend God except by acting contrary to the human good.[6] Hence Vatican I (1869–1870) and II (1962–1965) declared that the Catholic Church, because it is Apostolic, Catholic, One, and Holy, is a "moral miracle" calling for the consent of faith of those who become acquainted with it.[7]

In order to take faith in full seriousness, we might say, one is obliged *also* to take reason seriously, since the fullness of human good for our species depends upon both. It is not that one should try to "prove" faith by scientific means, for that is impossible and, as Aquinas says,[8] demeans the very faith involved. But it *is* important to show how the truths of faith, while lying "beyond" what reason *can* prove, yet are not *contradicted* by what science learns in the course of the advance of human understanding, and indeed are *enriched* thereby.

Truth cannot contradict truth: this formula we find enunciated over time, again and again in documents of the Church — and always on the same ground: that "it is the same God who reveals the mysteries and infuses faith, and who has

---

[6] Aquinas 1259/1265: *Summa contra gentiles* III, chap. 122: *"Non videtur esse responsio sufficiens si quis dicat quod facit iniuriam Deo. Non enim deus a nobis offenditur nisi ex eo quod contra nostrum bonum agimus."* See Chapter 4 below.

[7] See Ashley 2000: *Choosing a Worldview and Value System: An Ecumenical Apologetics*, pp. 211–244.

[8] Aquinas 1264: *On Reasons for Religious Belief* (*De Rationibus Fidei*), chap. 4 on "how to argue with unbelievers".

endowed the human mind with the light of reason".[9] So it is a question of *one truth* in eternity, indeed, but one truth made up of many truths in time, some of which are demonstrable by reason, while others of which are revealed by God in the Scriptures and developed in religious tradition. To develop and demonstrate not only the compatibilities, but further the *mutual enrichment* of the truths of faith and the truths attained by reason is the ongoing task of theology, toward which we offer this book as a contribution.

Benedict Ashley, O.P., and John Deely
23 July 2011

[9] Vatican Council I 1870: *Dei Filius*, Session 3, Chap. 4, pars. 5–6. See also *Catechism of the Catholic Church* #159.

CONTENTS IN DETAIL

HOW SCIENCE
ENRICHES THEOLOGY

Chapter 1

THE POSTMODERN DILEMMA OF THEOLOGY AND SCIENCE

Catholic Thinking Has Neglected Modern Science

Vatican II, in *The Church and the Modern World* (*Gaudium et Spes*), n. 5, marveled at the progress of modern science:

Today's spiritual agitation and the changing conditions of life are part of a broader and deeper revolution. As a result of the latter, intellectual formation is ever increasingly based on the mathematical and natural sciences and on those dealing with man himself, while in the practical order the technology which stems from these sciences takes on mounting importance. This scientific spirit has a new kind of impact on the cultural sphere and on modes of thought. Technology is now transforming the face of the earth, and is already trying to master outer space. To a certain extent, the human intellect is also broadening its dominion over time: over the past by means of historical knowledge; over the future, by the art of projecting and by planning.

In n. 36 of this document, the Council also rejected fundamentalist attacks on scientific research, when it said:[1]

---

[1] Vatican II, "The Church in the Modern World" (*Gaudium et Spes*), n.36. <http://www.vatican.va/archive/hist_councils/ii_vatican_council/docum ents/vat-ii_const_19651207_gaudium-et-spes_en.html>

We cannot but deplore attitudes (not unknown among Christians) deriving from a shortsighted view of the rightful autonomy of science; they have occasioned conflict and controversy and have misled many into opposing faith and science.

Yet the Council said little about what the remarkable advances of modern science might mean for theology. Nor have the Popes; although John Paul II apologized for the Church's treatment of Galileo and admitted the factual certitude of biological evolution.[2] No doubt this by-passing was due to the fact that the theological advisers of the conciliar bishops, although much interested in questions of history, ecumenism, and social justice, had not given much thought to the question. Thus one of the most prominent of these advisers, Karl Rahner, SJ, near the end of his life humbly admitted:[3]

> Each time I open some work of whatever modern science, I fall as theologian into no slight panic. The greater part of what stands written there I do not know, and usually I am not even in the position to understand more exactly what it is that I could be reading about. And so I feel as a theologian that I am somehow repudiated. The colorless abstraction and emptiness of my theological concepts frightens me. I say that the world has been created by God. But what is the world — about that I know virtually nothing and as a result the concept of creation also remains unusually empty. I say as a theologian that Jesus is as man also Lord of the whole creation. And then I read that the cosmos stretches for billions of light years, and then I ask myself, terrified, what the statement that I have just said really means. Paul still knew in which sphere of the cosmos he wanted to locate the angels; I do not.

---

[2] See Michael Sherwin, OP, 1996: "Reconciling Old Lovers, John Paul II on Science and Faith".

[3] Rahner 1984: "The Experiences of a Catholic Theologian", p. 412.

One might suppose that Protestant theologians with their emphasis on the fallibility of fallen human reason would be even more negligent with regard to this problem. Yet their need to defend the Bible as the sole source of their faith against Darwinism and the Big Bang has been more urgent than for Catholics for whom the Bible is interpreted in the light of Sacred Tradition. Protestants, at least in the United States, have taken the lead in promoting dialogue between religion and modern science.[4] Yet even such Protestant efforts do not venture much beyond attempts to answer attacks on religion by showing that science does not necessarily contradict faith.

One of these Protestant theologians, Ian Barbour in his excellent discussion, *When Science Meets Religion: Enemies, Strangers, or Partners,*[5] distinguishes four approaches to the relation of theology and science: (1) conflict, (2) independence, (3) dialogue, (4) integration. Of these four approaches, the second, *independence*, although it currently has influential advocates, seems to us (as Rahner lamented) the least feasible. In all religions some attempt is made to deal with questions about the causes of natural phenomena. Thus theology and science inevitably overlap and must either attack each other or find common ground. The first approach, *conflict*, while no doubt inevitable unless somehow reasonably resolved, is extremely damaging to the objectives of both theology and science. The third approach, *dialogue*, is a necessary first step,

---

[4] The magazine *Science and Theology* supplies information on these various developments. It is funded by the Protestant John Templeton Foundation, see http://www.templeton.org/. Also the magazine *Zygon: Journal of Religion & Science*, is a publication of the Institute on Religion in an Age of Science (IRAS), founded in 1954. IRAS is an independent society, affiliated to the American Association for the Advancement of Science and a member of the Council on the Study of Religion. It is also sponsored by The Zygon Center for Religion and Science established in 1988 as a partnership program of the Lutheran School of Theology at Chicago and the Center for Advanced Study in Religion and Science (CASIRAS).

[5] Barbour 2002.

but will yield only a *modus vivendi* unless it is engaged in with honesty and courage to seek an increasing area of real agreement and integration. In this book we will make an effort to do just that.

By this integrative approach to the relation of theology and modern science we emphatically do not mean that either discipline ought to abandon its own proper autonomy or methodology. Rather, the aim of an integrative approach should be to establish a *positive, systematic* relationship between theology and science, such as was attempted in medieval universities.[6] Surely theologians should exemplify a kind of intellectual humility by which they recognize that their own discipline can be enriched by scientific advance without distortion of theology's own proper method. Perhaps their example can encourage scientists also to adopt a similar critical openness. That this integrative task will not be an easy one is evident from certain past failures, such as the decline of Protestantism in recent years due to the confused debate of fundamentalists vs. liberals.

Thus, before attempting this integrative approach to science and theology, it is necessary in this introductory chapter to clear up four ambiguities that stand in the way of such an approach: (1) What is the relation of "science" to "the philosophy of science"; since it might seem that it is the philosophy of science, not science as such, which can enrich theology; (2) Ought we to "Dream of a Final Theory", since if this is the goal of science it would seem to replace, not enrich, theology; (3) How is the atheism that many scientists think has exposed theology as nonsense to be answered; (4) What is the goal of theology that opens it to scientific enrichment?

---

[6] The shortcomings of medieval science were due to the fact they had no scientific laboratories, advanced instruments of observation, or advanced mathematics. This deficiency began to be corrected only in the seventeenth century when the practical uses of scientifically based technology, such as Galileo's telescope, became evident and with them in the nineteenth century the increased societal investment of research funds. Scientific investigation is expensive, whence at some points it has to "pay off"!

Is There a "Philosophy of Science"?

Catholic theology was for a long time protected against the achievements of modern science by the Late-Scholastic and Neo-Scholastic distinction between "philosophy", which was supposed to deal with the "ultimate causes of the being of things", and "science", which was supposed to deal only with "phenomena", i.e., things falling within the realm of what can be seen and touched and "quantified" or measured. Even sound Thomistic scholars still use this "science vs. philosophy" distinction and its accompanying terminology as if it were that of St. Thomas. In fact, even such terms as "ontology" and "psychology" were unknown to Aquinas, but were the invention in late sixteenth and early seventeenth German universities of Protestant Scholastics who were not Thomists.[7]

The modern philosophical terms in question gained their popularity especially through the textbooks written by Christian Wolff (1679–1754), an Enlightenment Cartesian-Leibnizian idealist and rationalist.[8] When, in 1879, Pope Leo XIII initiated the Thomistic Revival, the history of Thomism was

---

[7] According to Kirstić 1994, "in a document known for years there is a detail which has unfortunately remained unnoticed until now and which fully entitles us to a complete revision of the established opinion on the first appearance of the word 'psychology' in the scientific language of Europe. At least 66 years before Gockel (and also a few years before the publication of Melanchton's lectures "on the soul"), the term 'psychology' was used by our great humanist, the poet of 'Judita', Marko Marulic (1450–1524) in one of his Latin treatises not as yet found but whose title 'Psichiologia de ratione animae humanae' is preserved in a list of Marulic's works given by the poet's fellow-citizen, contemporary, and friend Bozicevic-Natalis".

Nonetheless, it was through the work of Rudolf Göckel (1547–1628; sometimes referred to as "Glocenius", other times as "Goclenius"), and of Protestant Scholastics after him, this term "psychology", as also the term "ontology", came into the mainstream usage of modern philosophy. Cf. Raul Corazzon, *Theory and History of Ontology. A Resource Guide for Philosophers* at <http://www.formalontology.it/>.

[8] See Richard Blackwell 1961, "The Structure of Wolffian Philosophy"; also Matt Hettche 2008: "Christian Wolf", in the online *Stanford Encyclopedia of Philosophy* <http://plato.stanford.edu/entries/wolff-christian/>.

quite poorly known, with the result that this distorting terminology has been unfortunately retained in the Neothomistic
(or late modern Thomistic) presentations of Aquinas' thought,
even by excellent historians of Medieval Philosophy such as
Étienne Gilson.

Why "unfortunately"? Because, as John Deely has noted,[9]

> Epistemology and ontology, but especially *epistemology*, is
> "an offspring of philosophical modernity" in ... the sense in
> which philosophical modernity goes the Kantian route of
> severing "*things*" (what exists *often* prior to, but *always* inde-
> pendently of, our mental representations, whether self-rep-
> resentations or other-representations), from "*knowability*".

Fr. Benedict Ashley, for his part, has tried to resolve this
dilemma in detail in a lengthy book *The Way Toward Wisdom*,[10]
along a line of thinking that can be stated here briefly as follows.
Modern thought since Descartes at the time of Galileo and
Poinsot, while vehemently claiming to be empirical, has in fact
based knowledge on Platonic ideas or, since the Enlightenment
of the 18th century, on the innate "logical" and a-priori categories of Immanuel Kant. If we are to be really empirical, we must,
following Aristotle and Aquinas, reject this idealism and rationalism and base all our critical knowledge on the evidence of
the senses, as scientists certainly strive to do, but recognizing

---

[9] "Realism and Epistemology" (Deely 2010: 80),

[10] Ashley 2006, *The Way Toward Wisdom: An Interdisciplinary and Intercultural Introduction to Metaphysics*. In this work Ashley also shows that the First
Way of proving the existence of God, which is often said to be "metaphysical", pertains rather properly (as Aquinas himself says) to natural science, and
is *presupposed* to the validity of metaphysics as a discipline. What metaphysics
adds to it is to show that it holds not only for our universe as natural science
proves it to be, but also for *any possible* universe, which is why this "proof
from motion" is the most basic of all humanly formulable proofs of the divine existence — why it is, as St. Thomas put it, the *prima et manifestior via*
— "the first and and most manifest way". See now the comparable discussion
in Deely, *Medieval Philosophy Redefined* (2010b: 187–191), "The Reasoning of
the Five Ways".

the while that this critical control of objectification[11] in which science most centrally consists — the process of making things known on an intellectual and reflexively aware basis — cannot be simply reduced within human understanding to what can be sensed directly, as the Enlightenment idea of science came to hold and as we see surviving today in the crippled notion of "science" advocated by the "four horsemen" of the "new atheism",[12] as also most recently by Stephen Hawking.[13] Aquinas called the critical discipline which first attempts to exercise this species-

---

[11] The expression comes from Fr. Ralph Austin Powell. Cf., for example, Powell 1982. His general idea was that philosophical science and science in the modern, empirical and mathematical sense, share a common framework rooted in the *human* use of signs, that is to say, not merely in the use of signs as an activity that is common to all animals, but in the use of signs based on the species-specfically human awareness of the *difference* between objectification and the world of things as being more than and sometimes at odds with how those "things" are transformed into objects through use of signs. The process of objectification, precisely, in order to reach beyond the level of animal practicality to the level of science, has to be *critically controlled*. And this process of critical control is not only central to science in the modern sense, but is also *presupposed* to the very possibility of modern science as a coming to terms with things in their physical, and not merely objective, being. Charles Peirce will thus distinguish between science as *idioscopic*, or based on the use of instruments as extending our senses, and *cenoscopic*, or resulting from a critical control of objectification that makes the development and use of instruments possible in the first place, as discussed in Ashley 2006. Deely 2008: *Descartes & Poinsot: the crossroad of signs and ideas*, taking up Peirce's terminology on this point, will argue that a postmodern development of intellectual culture is best served by understanding the contrast between scientific knowledge and philosophical knowledge in these terms; for while both cenoscopic and idioscopic knowledge are, as critically controlled objectification, *scientific* knowledge available as such only to human animals, this is a distinction which also makes clear the inescapable priority of cenoscopy over idioscopy.

[12] I.e., Sam Harris (1967–), Richard Dawkins (26 March 1941–), Daniel Dennett (28 March 1942–), and Christopher Hitchens (13 April 1949–). Not that these four by any means exhaust the contemporary "atheism" movement as a postmodern phenomenon dependent upon the Enlightenment view of science as directly and exclusively concerned with material substance (e.g., add in Victor Stenger [29 January 1935–], et al.); but the analogy is to the "four horsemen of the Apocalypse".

[13] See Hawking and Mlodinow 2010.

specifically human critical control of objectification *scientia naturalis* or, using Aristotle's term, *physica*.

For Aquinas all critical, logically systematic disciplines are "sciences" in Powell's sense of critically controlled objectifications,[14] so that mathematics and political theory are also "sciences". Moreover, for him each and all such systematic "sciences" are "philosophy" in the sense of searches for wisdom. The term "metaphysics" (after-physics), which Aristotle himself never used, means a discipline that relates and criticizes all the other kinds of philosophy (the special sciences) while preserving their *autonomy*. This concern for the autonomy of the different sciences (the division and order of the sciences) was one of the main points on which Aristotle had corrected his master Plato, who had come to suppose that all human knowledge can be reduced to a single supreme Idea. Aquinas agreed with Aristotle that metaphysics — the science of being as common to the finite order of existence as consisting of more than only material beings — has no data of its own, but only has the task of analyzing and relating the data of the other sciences as revelative of the full extent of "what exists". Hence, although metaphysics is the supreme science in the sense that it concerns the very foundation or what is *First* in the order of learning and proof; yet as to the data on which it reflects, it presupposes what other science have brought to light (or "objectified"), and hence is the *Last* to be known.[15]

The sciences other than natural science are all more abstract than is natural science, for natural science or "physics" is what deals most directly with the sensible order of being as material. Hence the other sciences in this sense, just as does metaphysics, all *presuppose* natural science for their own validity. Do they, however, presuppose *all* of the conclusions of natural science? By no means, since our exploration of nature

---

[14] Note 11 above.

[15] See "Why First Philosophy is Last", in Ashley 2006a: *The Ashley Reader: Redeeming Reason*, pp. 27–46.

will inevitably always be incomplete. What the more refined truths of scientific research, gained by experiments and the use of instruments, presuppose are only those basic truths of natural science that are most directly evident to us from an analysis of our daily common sense experiences. It is these basic questions that today are lumped into a "philosophy of science" and often assigned to metaphysics; but they do not pertain to any other discipline than natural science itself, but are its first principles, just as the axioms and basic theorems of mathematics pertain to mathematics itself, not to a "philosophy of mathematics", and are essential for its other findings to be critical and systematic.

Unfortunately, as we will show next, in modern scientific education, very little attention is given to this analysis of physical principles.[16] Instead, the student is immediately plunged into the use of mathematical constructs. Since mathematics is more abstract and remote from sense experience than is natural science, this raises a problem about the certitude of scientific conclusions that depend in part on mathematics. Our observation of nature, even when refined by experiments and instruments, is never precise in the way a mathematical construct is. What is mathematically impossible is physically impossible, but not the reverse. Thus only negative mathematical demonstrations can give physical certitude, while positive ones can only yield probability. Hence mathematical constructs, like the astronomies of Ptolemy, Copernicus, Newton, and even that of Einstein, are only probable. Some day current Einsteinian astronomy will probably have to be replaced unless we can interpret it realistically so that its proofs are based on the principles of natural science founded directly on sense experience, rather than mainly on the principles of mathematics.

---

[16] Thus, for example, when we apply Peirce's terminology as in Deely 2008, we are forced to realize how little attention is given to the dependency of ideoscopic science for its validity upon cenoscopic science as providing the prior framework and ground for experimental and mathematics applications.

The most evident and certain information given us by our senses is that *all that we sense is moving*, and it is precisely this fact that is basic to all the rest of science. If the world were not in motion we could not sense it, since it must move into contact with our body directly or through a changing medium in order for us to be able to sense it. To describe any observed object we must describe both its present actuality and, in order to do this (since the present itself never ceases to draw on a future), some measure of the *future* states into which the observed object is passing. To do this, as long ago Aristotle showed in his *Physics*, requires us to observe its nine generic properties, its *quantity* (or spatial materiality), its *qualities* (as arising from its form as this rather than that sort or species of being), its *actions* (upon what surrounds it) and *receptions* (of action from the surroundings), its *relations* (as consequent upon actions and receptions, the "children" thereof, as it were, making forensics possible),[17] its *place* (or "where", i.e., its location among surrounding bodies), its *posture* (or arrangement of its own parts in place, e.g., standing, squatting, sitting, lying, etc.), its *time* (or "when" in the before and after of environmental transitions as measured against more regular changes), and its proximate and immediate *environment* (the more remote object through which the moving object will pass, such as the water in which a fish swims).

------

[17] Note in particular that, as first pointed out in Deely 2001: 73–77, the categories posterior to relation as listed in this way themselves involve relations. That is to say, 'substance', 'quantity', 'quality', 'action', and 'passion' precede and are not themselves 'relation', but they give rise to and support relations; then, in order to understand the remaining four categories of 'where', 'when', 'posture', and 'vestition', relation is presupposed. Whence if relation itself, as Ockham and the moderns later hypothesized, were not *itself* realizable in the order of mind-independent being, then neither could the remaining four categories truly be classifications within the order of mind-independent being (*ens reale*) conceived as unmixed with mind-dependent being. Of course, the most serious consequent of denying *ens reale* status to relation is that it makes the world beyond mental representations unreachable in itself, exactly as Descartes discovered and Kant systematized.

From this analysis of how bodies change three principles are evident: (1) the Principle of Non-Contradiction, that as a thing changes it cannot *simultaneously* but *only sequentially* be in opposite states; (2) the Principle of Causality, that a thing cannot change itself in the very respect that it changes, since it cannot give itself an actuality it does not already have; (3) the Principle of Research, that one must proceed from what is better known to what is less well known. Current science uses these principles, but seldom analyzes them in depth, with the result that scientists seldom provide a critical physical interpretation of their mathematical constructions. Is a "wave particle" a wave or a particle, or what is it really like? We do understand why the earth is round and not flat; but, as a Nobel Laureate in physics, Richard P. Feynman, said about the general situation in physics today:[18]

> What I am going to tell you about is what we teach our physics students in the third or fourth year of graduate school.... It is my task to convince you not to turn away because you don't understand it. You see my physics students don't understand it .... That is because I don't understand it. Nobody does.

Dreams of a Final Theory

What then do modern scientists think they are achieving? Another Nobel Laureate in Physics, Steven Weinberg (1933–), has dealt with this in his 1992 *Dreams of a Final Theory: The Search for the Fundamental Laws of Nature*. At present modern science rests on two fundamental theories that taken together are called the Standard Theory: Einstein's Special and General Theories of Relativity dealing with macro-physics on the one hand, and on the other Quantum Theory dealing with microphysics. The "dream" is to develop a Unified Theory that will put together the four fundamental forces of gravitation,

---

[18] Feynman 1985: *QED, The Strange Theory of Light and Matter.*

electromagnetism, and the weak and strong nuclear forces. The most promising proposal for a solution for this problem is String Theory, which we will discuss briefly in Chapter 2, but which to date remains neither verified nor falsified. Hence books such as David Lindley's 1993 *The End of Physics* and John Horgan's 1996 *The End of Science* are beginning to appear.

Weinberg, however, is confident that eventually such a Final Theory will in fact be achieved. He insists, moreover, that this will not mean the end of scientific research, since such a theory will be only generic, while the laws governing specific situations and types of change will still indefinitely remain to be explored. For example, we are still far from perfectly reducing psychology, biology and even chemistry to physics. It is also to be noted that the Big Bang Theory and Evolution are more and more historical rather than law-like theories.

Theological enrichment therefore must come both from what is certainly known and from the *direction* suggested by these dialectical constructions; but the two must be carefully distinguished. Weinberg believes, however, that when his dreams come true this Final Theory will render the universe self-explanatory, and thus eliminate once for all a belief in the existence of a Creator God — much as Hawking claims in his latest book.[19] In part Weinberg's confidence is based on the Multi-Universe Theory of Andrei Linde and Martin Rees,[20] which argues that since many or even an infinity of possible universes may exist, it is only an accident that the one we live in it has such puzzling laws. Thus Weinberg writes:[21]

> About the multiverse, it is appropriate to keep an open mind, and opinions among scientists differ widely. In the

---

[19] Hawking and Mlodinow 2010.

[20] See Linde 1990: *Particle Physics and Inflationary Cosmology*. The British Sir Martin Rees, however, in his 2004 *Our Final Century: Will the Human Race Survive the Twenty-first Century?*, argues that we may not survive modern technology.

[21] Weinberg 2005: 40, "Living in the Multiverse".

Austin airport on the way to this meeting I noticed for sale the October issue of a magazine called *Astronomy*, having on the cover the headline "Why You Live in Multiple Universes". Inside I found a report of a discussion at a conference at Stanford, at which Martin Rees said that he was sufficiently confident about the multiverse to bet his dog's life on it, while Andrei Linde said he would bet his own life. As for me, I have just enough confidence about the multiverse to bet on it the lives of both Andrei Linde's and Martin Rees's dogs.

This does not mean that the Standard Theory (or a Final Unified Theory) is nonscientific. In fact it is backed up with extensive and precise empirical evidence. Instead, it means that — like Ptolemy's astronomy — the Standard Theory is at best only a mathematical construct that touches reality at certain points but remains in large part conjectural and difficult to interpret in real physical terms. Indeed, if these many universes are unconnected so that the others have no effect on the one we are in, we can never know their existence or anything about them. On the other hand, if they influence ours they are simply regions of a single universe and, as such, merely add to our difficulty in understanding our own universe.

Can Natural Science Logically Be Atheistic?

Like Einstein, Weinberg admits by his hope for a Final Theory that the universe has a first cause, but assumes it is simply some material body the actions of which are predicable by a mathematical formula that is without initial conditions, and hence need not be a personal God. He writes:[22]

> Either you mean something definite by a God, a designer, or you don't. If you don't, then what are we talking about? If you do mean something definite by 'God' or 'design', if for instance you believe in a God who is jealous, or

---

[22] Weinberg 1999: "A Designer Universe?".

loving, or intelligent, or whimsical, then you still must
confront the question 'why?' A religion may assert that
the universe is governed by that sort of God, rather than
some other sort of God, and it may offer evidence for this
belief, but it cannot explain why this should be so. In this
respect, it seems to me that physics is in a better position
to give us a partly satisfying explanation of the world than
religion can ever be, because although physicists won't
be able to explain why the laws of nature are what they
are and not something completely different, at least we
may be able to explain why they are not slightly different.
For instance, no one has been able to think of a logically
consistent alternative to quantum mechanics that is only
slightly different. Once you start trying to make small
changes in quantum mechanics, you get into theories
with negative probabilities or other logical absurdities.
When you combine quantum mechanics with relativity
you increase its logical fragility. You find that unless you
arrange the theory in just the right way you get nonsense,
like effects preceding causes, or infinite probabilities. Religious theories, on the other hand, seem to be infinitely
flexible, with nothing to prevent the invention of deities
of any conceivable sort.

Weinberg ignores the fact that the pagan Aristotle in his
Physics Book VIII, and Aquinas in his First Way in the *Summa
Theologiae*, I, q. 2, art. 3, argued for the existence of God not by
the Design Argument which Plato had used, nor through a "philosophy of science", but by a strictly empirical proof as part of
the basis of the whole of physics. When we see something moving we know from experience that this motion is an *effect* that
must have a *cause*. This primary and best known fact of change
is *the only fact required to prove that change is ultimately the effect of
an absolutely unmoved mover*[23] which, since it causes that effect
to exist (Principle of Causality), must also itself exist (Principle

---

[23] See the discussion of this point in note 10 above.

of Non-Contradiction[24]). This cannot be an infinite chain of moved movers, because in such a chain every mover would only be potentially moved and hence cannot move the next mover or cause the final observed motion. Thus the existence of the whole universe of moving things studied by science depends on the existence of a first unmoved mover here and now, not at some past moment. Moreover, since all material things are subject to change, it is evident that this unmoved mover is not material.[25]

Or, to put this proof somewhat differently, moving things that may at first sight seem to be moving themselves, on examination prove always to be moved by another, and the inquisitive scientist seeks that cause. Living things seem to move themselves, but the biologist researches to find out from what they get their energy — from food, air, sunlight, etc. The fundamental particles from which all bodies are composed are in constant motion, and physicists strive to find the source of their energy. Today this is traced to the fundamental forces of gravity, electromagnetism, and the weak and strong nuclear forces. Weinberg is confident that eventually these natural forces will be shown to derive from one single force that will have "no initial conditions". Yet, since these fundamental forces are not abstract mathematical entities but properties of bodies which they have received and then exercise through change, they are moved movers and cannot be self-explanatory. Thus atheism is a position that undercuts the very foundations of natural science, since it contradicts the Principle of Causality. (Of course, if such a contradiction could be the case, ironically, the very possibility of finding the natural causes of the facts we observe would be in every case eliminated!).

---

[24] Also called, with meaning unchanged, (one of the many ironies in philosophy's history) "the Principle of Contradiction": see Deely 2001: 121 note 77 and 356.

[25] This approach is exemplified by the distinguished physicist Anthony Rizzi in his 2004 book *The Science Before Science*, as also in his 2008 *Physics for Realists*; and in the work of his Institute for Advanced Physics <http://www.iapweb.org/>.

That is why David Hume (1711–1776), who mistakenly thought that effects simply follow their causes but are not dependent on them, became a skeptic. Hume's skepticism forced Immanuel Kant (1714–1804) into idealism and the denial that we know material things in themselves (*Dingen an sich*). In turn, Kant forced Auguste Comte 1798–1857 into Positivism, the reduction of science to a mere description without causal explanation. It is sad, therefore, that any scientist should contradict his work and himself by claiming to be an atheist or agnostic, or even, as Newton did, a deist, or, as some have taken Einstein to be, a Spinozistic pantheist.[26]

Yet many authors with less authority than a Nobel Laureate in Physics such as Weinberg are now producing popular books advocating atheism. There is Daniel C. Dennett's 2006 book, *Breaking the Spell: Religion as a Natural Phenomenon*. Richard Dawkins, also in 2006, offers his violently rhetorical book, *The God Delusion*. More violent still is the 2007 work of Christopher Hitchens (not a scientist but a political writer), *God is Not Great: How Religion Poisons Everything*. Perhaps most serious of all is the 2007 work of the physicist Victor J. Stenger, *God: The Failed Hypothesis*. Stenger ends by calculating as positive the probabilities that the universe emerged from empty space, that is from "nothing"! "Nothing", however, explains nothing.

The current campaign of some Christians to answer such scientific arguments for atheism by the Anthropic Cosmological Principle or Design Argument has had, as mentioned before, little success because it appeals to complicated mathematical data that can be questioned. It must be raised, as we have already indicated, at the much more fundamental level

---

[26] Max Jammer (1999: *Einstein and Religion*, p. 48), however, reports the following remark: "I'm not an atheist, and I don't think I can call myself a pantheist. ... I am fascinated by Spinoza's pantheism, but admire even more his contribution to modern thought because he is the first philosopher to deal with the soul and body as one, and not two separate things."

of science *presupposed* to all mathematical hypotheses, namely, with the fact of change and motion.[27]

Furthermore, there is a glaring logical contradiction in supposing, as these atheistic scientists do, that the discovery of some Final Theory will locate the ultimate cause of change in a material body or bodies having an ultimate force and thus, unlike the previous forces in the chain, *self-moving*. Material causes are bodies that are changeable and hence, although they actually exist in some particular state, are *potentially* some other kind of thing or are at least potentially in some other state, or are actually moving into that state. "They cannot give what they do not have", and thus cannot give themselves the actuality to which they are only potential. Nor, as we have argued above, can an infinitely long chain of moved movers produce an observed motion because, without a first unmoved mover as their cause none of them can actively produce the observed effect.[28]

While it is true that this classical First Way of proving God's existence only shows the existence of a First Cause defined as *not material*, from this conclusion follow further arguments that, by analogy from this First Cause's observable effects, show its nature.[29] This Uncaused Cause must have in a transcendent way all the perfections that it has freely given to

---

[27] In the presentation by Aquinas of "five ways", the Second and Third Ways are closely related to the First Way, but use different middle terms. While the First Way proceeds from the most evident of facts, namely *motion*, the Second proceeds from the existence of agents of motion and the Third from the fact of contingency, that things that move may sometimes not exist.

[28] See, for a detailed discussion of the First Way, *The Way Toward Wisdom* (Ashley 2006), pp. 92 –131; and *Medieval Philosophy Redefined* (Deely 2010b), pp. 187–191. More will be said about this during the course of the present book. The usual objection to this proof is based on Kant's attempt to refute it by confusing it with the a-priori "Ontological Argument" of St. Anselm, an argument that Aquinas explicitly rejects (1266: *Summa theologiae* I, q. 2, art. 2). Instead, Aquinas gives, in art. 3, strictly *a posteriori* proofs. On this see O'Brien 1960: *Metaphysics and the Existence of God.*

[29] See the development in Deely 2001: 267–290; and 2010b: 187–207.

its freely created universe. Since this Uncaused Cause created us human beings, it must be intelligent and free, and thus *personal* or, rather, a Super-Person — truly God. Thus Weinberg is mistaken in saying that reason cannot explain why God is not stupid and without free will but intelligent and free, since Weinberg, one of God's more interesting creations, although by choice closed-minded to the classical arguments for God, surely is very intelligent and free.

Therefore science rests entirely on foundations that include proof of the existence of God as Creator without whose creative and sustaining action the universe studied by science could neither exist nor change, nor could human scientists hope to understand and control it. Sound science is consistent only with monotheism, and exposes the error of all forms of monism. A materialist monism is impossible, because matter would not exist without a non-material, spiritual First Cause. A spiritualist monism, like that of Hindu mysticism,[30] is equally impossible, because it treats the sensible world as merely phenomenal (as did also Plato), and thus deprives bodily human beings of any trustworthy way of coming to know a spiritual God.

Thus the first way in which modern science enriches Christian theology is that it exposes the monism of many of the world's religions, including the Secularism of many of today's scientists who are afraid to admit the existence of a Creator of

---

[30] Recently, Francis X. Clooney, S. J., an expert on Hindu texts, which the present authors are not, has contended in his 2005 *Divine Mother, Blessed Mother*, that Hindu thought is monotheistic (Clooney 2005: 111, 135), although its supreme God is represented as having co-relative male-female gender. Yet he does not show that any of the many forms of Hindu religion or philosophy hold for *creatio ex nihilo*, that is, that God has *freely* created the material and spiritual universe having true existence in itself yet an existence totally dependent on Him, while He remains totally free of the created universe. Other scholars elaborate on Hindu beliefs as ranging from non-dualist pantheism (spiritual or material) through polytheism to atheism; yet, apart from the influence of Judaism, Christianity, or Islam, none of these scholars brings to light a Hindu teaching that advocates true monotheism. See Smart 1999, *World Philosophies*, and Küng et al. 1993, *Christianity and World Religions: Paths to Dialogue*.

the universe from nothing — a position contradictory to the very foundations of science and sound scientific method.

But is it hopeless to think that science can be built on empirically evident principles? Postmodern critics of what they call "foundationalism" claim that in fact scientists only seek empirical facts in view of some theory they *want* to prove, and thus as scientific research proceeds it is more and more built up on an infinite set of hypothetical presumptions.[31] This postmodern objection does at least expose the fact that, since the time of the French Revolution, secularism has assumed on the basis of a few scandals like the Galileo Affair that the monotheism, taught by Judaism, Christianity, and Islam, which attributes the universe to a Creator, is the enemy of science.

The notions of a "creator" and "creation", however, are used in two quite different senses today: (a) *monotheistic* religions teach that One God exists in total independence of the universe he creates, which, although it truly exists, totally depends on its Creator for that existence: (b) *monistic* religions teach that all reality is really one Absolute Existent, whether this be conceived as entirely material, or inclusive of spiritual elements, or wholly spiritual. Today the world religions, roughly grouped under the widest possible definitions, have the following percentages of adherents:[32] 33% Christian (Catholic 53%, Orthodox 11%, the remaining 36% Protestant and others), 21% Islam, Secularist 16%, Hindu, 14%, Buddhism 4%, Chinese Traditional, 4%, Primal Indigenous, 4%, Other 4%, with Judaism a mere 0.22%. Thus Christianity and

---

[31] This is argued exhaustively Michel Foucault 1970, *The Order of Things: An Archaeology of the Human Sciences*, although he has little to say about natural science.

[32] See "Major Religions Ranked by Size", <http://www.adherents.com/Religions_By_Adherents.html> and <http://netvangelize.org/8359>. This last site notes that these are rough, round estimates, and hence we have reduced each of *Buddhism, Chinese Traditional, Primal Indigenous, Other* from 6% to 4% to get 100%.

Islam constitute more than half of the world's present population, and have a worldview and value-system derived from
Judaism, although less than 1% of about 14 million Jews are
now Orthodox. What these three religions have most clearly
in common, however, is their strict monotheism.[33] The other
religions are essentially *monist*. Indigenous religions that are
expressed principally in mythological, not systematic, philosophical terms are so vague as to incline to monism, although
they may imply a certain original monotheism by their worship of a "High God".[34]

The various forms of Hinduism, Buddhism, Taoism, and
Confucianism are not simply mythological, but philosophically systematized; yet they tend to a mystical spiritual monism
in which the reality of the universe and even the individual
human spirit is rejected. As such they can be classed with Platonism, especially in its Neo-Platonic form that underlies
Western culture and may even have been influenced by Indian
thought. Jacques Maritain in a brilliant essay, "The Natural
Mystical Experience and the Void",[35] explained in Thomistic
terms how this experience can be genuine and may even be
elevated by grace. Yet in its natural obscurity such experience
tends to a monistic interpretation. Spiritual monism, as the
cultures where it has predominated show by their failure to
develop modern science (for which they depend entirely on
the Christian or post-Christian West), essentially discourages

---

[33] Jews and Muslims reject Christian Trinitarianism as inconsistent with
strict monotheism, but this is a misunderstanding, inasmuch as the Three
Divine Persons have a single divine existence. For discussion see Ashley
2000, *Choosing a Worldview and Value System*, pp. 123–127 and 136–140.

[34] Ancient Zoroastrianism, which may have had some influence on Judaism, was dualistic, teaching the existence of both Good and Evil principles.
It fostered Gnosticism, Manichaeism, and other cults now almost extinct, yet
its dualism reduced to monism because it held that the universe is the work
of the Evil God who, with all his works, will eventually be extinguished by
the true God.

[35] Maritain 1943.

scientific research in favor of a search for the Absolute through mystical introspection.

Ninian Smart rightly surveys all these world views under the title of his 1999 book, *World Philosophies*, because only the three monotheistic religions emphasize faith in a *revelation* from the Creator that transcends truths accessible to human effort, since the monistic world views are based not on faith in a revelation but on natural human experience as attainable at least by adepts. Neither Siddhartha Gautama (563–483 BCE), the historic Buddha, nor Shankara, (788?–820? CE, often called the Thomas Aquinas of Advaita Hinduism), nor the sage Confucius (551–479 BCE) claimed a revelation. Nor did they ask others for absolute faith in what they taught, but urged rather others to seek the same experiences that they enjoyed by the practices of virtue and introspection.

Among the three great monotheistic religions, Christianity places a special emphasis on what it calls (along with hope and charity) the theological virtue of *faith*. Faith is a virtue, precisely in that it is *reasonable*; yet it is *theological* rather than philosophical inasmuch as what faith accepts objectively transcends what reason can demonstate,[36] so that faith without God's grace transcends human possibility.[37] While the other monotheisms, Judaism and Islam, hold for faith in the One God, Christianity holds further for faith in the Trinitarian nature of God and the Incarnation as truths that, even though indemonstrable in themselves (i.e., inaccessible to rational proof positive), are yet *demonstrably not contrary* to the same reason which can prove God's existence, wisdom, and unity from his natural effects.

Secularism or Naturalism, now tending to dominate the post-Judaic-Christian West and which Weinberg typifies, is a

---

[36] For details in Aquinas on this matter, see in particular Deely 2010c: 279–301, "Projecting into Postmodernity Aquinas on Faith and Reason".

[37] *Catechism of the Catholic Church*, 1997: #s 26, 142, 150, 1814, 2087; John Paul II, 1998: *Encyclical On the Relation of Faith and Reason* (*Fides et Ratio*).

form of monism but, unlike most other monist doctrines, is materialist. It today claims only about 6% of humanity, but is increasingly influential globally, especially among the intelligentsia of many cultures. It puts its faith in the world-view provided by modern science and bases its value system on personal preference or communal relativism. Unquestionably modern science has made wonderful advances since the Scientific Revolution of the seventeenth century in explaining natural phenomena and increasing human control over nature, as for example, in reducing early deaths. Yet in fact the leaders of the Scientific Revolution, such as Copernicus, Galileo, Gassendi, Robert Boyle, and Newton, as well as many subsequent important contributors to scientific progress, such as Faraday, Mendel, Pasteur, Maxwell, Kelvin and Planck were sincere Christians. Thus Isaac Newton, a Unitarian Christian, in a letter to Richard Bentley dated 10 December 1692, said:

> When I wrote my Treatise [i.e., his great *Philosophiae Naturalis Principia Mathematica*, 1687] about our System [i.e., the universe], I had an eye on such principles as might work with considering men for the belief of a Deity; and nothing can rejoice me more than to find it useful for that purpose.

The founder of modern chemistry and an outstanding experimentalist, Robert Boyle (1627-1691), was also a vigorous Christian apologist,[38] who wrote:

> Nor was it his [God's] Indigence [lack of power], that forc't him to make the World, thereby to make new Acquisitions, but his Goodnesse, that prest him to manifest, and to impart his Glory; and the goods, which he so over-flowingly abounds with, Witness his Suspension of the World's Creation, which certainly had had an earlier

---

[38] On this, see the fine article of MacIntosh and Anstey 2010, "Robert Boyle" in the *Stanford Encyclopedia of Philosophy* <http://plato.stanford.edu/entries/boyle/>, whence is taken the quotation.

Date, were the Deity capable of Want, and the Creatures of Supplying it.

The current objections to a rational proof of God's existence confirm Aquinas' contention that there are only two serious arguments for atheism:[39] (1) that the universe requires no cause besides itself, which is Weinberg's view but which, as we have seen, involves a contradiction; and (2) the problem of evil: "If there is a good God, why is there evil in the world?" We will say more about this second argument in later chapters; but here let us only remark that even if one cannot answer this argument, the proof of a good God's existence remains valid, and like poor Job (42: 2–6) we ought to say to Him:

> I know that you can do all things, and that no purpose of yours can be hindered. I have dealt with great things that I do not understand; things too wonderful for me, which I cannot know. I had heard of you by word of mouth, but now my eye has seen you. Therefore I disown what I have said, and repent in dust and ashes.

In arguing for atheism, Weinberg refers to the First Way of Aquinas for proving God's existence, but dismisses it on the grounds that it is harder to explain why God exists than to explain to a high degree of probability why our universe exists of itself. This misses the point that, as Aquinas shows and we have expounded above, what is demonstrated by this First Way through the most primitive and certain facts of experience is that an Unmoved Mover exists as the cause of the changes we observe in our world. While it is contradictory to claim that physical things that exist in change cause themselves, no

---

[39] Aquinas 1266: *Summa theologiae* I, q. 2, art. 3, reply to objections 1 and 2. Recent authors (see Martin and Monnier eds. 2003) have tried to formulate a third argument for atheism, based on the claim that some attributes classically attributed to God are uninstantiable. We suspect that Aquinas would see in these "arguments" a radical misunderstanding of the way in which human language is applicable to a being in which existence is identical with essence. Cf. Deely 2001: 267–297, esp. 272–284; or 2010b: 191ff.

contradiction can be found in the concept of an Unmoved Mover that is non-material. Only after we are sure this non-material First Cause exists and causes all changing things, does it then become certainly evident that it is the super-personal God who is the cause of intelligent and personal scientists like Weinberg.

On the other hand, Weinberg, when asked what his research had taught him about the meaning of the universe, gave the notorious answer, "The universe is without meaning, it is pointless". When asked why he then continues his researches, he answered, "Because I enjoy the search". In *Dreams of a Final Theory* he elaborates this point by talking about "beauty", but not so much the beauty of the universe itself as of the mathematics of its current theoretical model! If he believed in God the Creator this enjoyment of beauty would not be diminished but greatly enhanced. As Psalm 19: 1 says, "The heavens declare the glory of God, the skies proclaim the work of his hands".

At the beginning of the Scientific Revolution when the religious wars of Catholics vs. Protestants and in England between Protestant churches led many intellectuals into skepticism, the great mathematician René Descartes struggled with the question whether we can be certain about anything. He thought he could find an answer based on the impossibility for a non-existent intelligence even to doubt. "I think, therefore I am", *Cogito ergo sum.* Cornelio Fabro, in his extensively documented 1968 book, *God in Exile: Modern Atheism; a study of the internal dynamic of modern atheism, from its roots in the Cartesian cogito to the present day*[40] has shown how this Cartesian idealism became the source of modern science's excessive reliance on mathematics that has led, contrary to Descartes' own intention, to modern atheism. This fatal tendency emerged when Benedict Spinoza proposed an explicit monism in which God and the universe were merely

---

[40] We do not agree, however, with Fabro's Wolffian distinction between "science" and "philosophy", for the reasons explained above in discussing "Is There a Philosophy of Science" (p. 7ff.; see esp. note 11, p. 9 above).

two aspects of a single substantial reality. It became even clearer when Immanuel Kant, still within the Cartesian tradition,[41] argued that reason cannot know God, nor can it even know the human conscious ego, but can only suppose their existence in order to maintain social moral standards.

Then Hegel argued that all of history is God's discovery of himself. Finally Karl Marx, as he claimed, "set Hegel on his head" by arguing that in fact all of history is simply the evolution of a universe of matter. Weinberg and most scientists do not accept Marxism as a social theory, but they accept a similar materialist monism. Thus the greatest of modern scientists, Albert Einstein, in answering frequent questions put to him about his religious views, replied that he could accept the "God of Spinoza" because Spinoza (1632–1670) was a mathematically-minded Cartesian but not a "personal God of fear".[42] When, therefore, Weinberg, concludes that there is no evidence for the existence of One God, the Creator, he does so because he is looking for the evidence at the wrong level and in the wrong dimension of the scientific enterprise.

Yet really it seems that the chief reason that many scientists, such as Weinberg, are so opposed to theism is that they fear it makes scientific research futile by explaining everything by a single explanation.[43] To believe that God is the First Cause of the universe, however, does not tell us what secondary causes He freely chooses to allow to develop and produce the effects

---

[41] Kant was educated in the philosophy of Leibniz who based his thought on the Cartesian proposal for a "new beginning" in philosophy and, although Kant came to attack Leibniz' view "critically", he in fact deepened its essential idealism. See "Immanuel Kant: the Synthesis of Rationalism and Empiricism", in Deely 2001: 553–570.

[42] See esp. Einstein 1930, 1941, 1948; overview in Holton 2002: "Einstein's Third Paradise"; see also Gilmore 1997. Re Spinoza's complexity, see Goldstein 2006 (reviewed in Bloom 2006).

[43] A similar misreading of Aquinas' "proof from motion" is evidenced in Hannam's otherwise excellent book of 2009, *God's Philosophers: How the Medieval World Laid the Foundations of Modern Science.*

we observe. In fact, monotheists also want to know more about these "secondary causes", not indeed in order to know that God exists, but, just as we would like to know more about Shakespeare through his dramas, to know more about God through his creation and, indeed, to really "get to know" that creation in the first place! Historically, this was an important motive for the work of the founders of the Scientific Revolution. As St. Paul said to the Romans (1:19-20): "For what can be known about God is evident to them [the Gentiles], because God made it evident to them. Ever since the creation of the world, his invisible attributes of eternal power and divinity have been able to be understood and perceived in what he has made".

Enriching the Trinitarian Economy

At the beginning of this 21st century Christian theology has at least three pressing tasks: *first*, theology must *preserve* the Sacred Tradition of the Christian Church, the *depositum fidei*; *second*, it must appropriately and apologetically (in the technical sense)[44] *interpret* this *tradition* to our world of the twenty-first century that is globally dominated by modern science and its technological applications; and *third*, to accomplish this second task, theology needs to *enrich its presentation of the Faith by assimilating the truths of modern science* to the context of Faith.

The Church Fathers assimilated the Faith, originally Jewish in expression, to the literary and philosophical forms of contemporary Hellenism.[45] The medieval Scholastic theologians

---

[44] "Apologetics" is that aspect of theology that undertakes to make the truths of revelation rationally credible to non-believers; see Ashley 2000, *Choosing a Worldview and Value System*; and Deely 2010b: 279–301, "Projecting into Postmodernity Aquinas on Faith and Reason", also cited above. Karl Rahner's theory of the "anonymous Christian", as argued in his 1978 *Foundations of Christian Faith* introduction to Christianity has led to the replacement of classical apologetics by courses in "Foundations of Theology"; but in Fr. Ashley's view this is a substitution that does not adequately meet the needs of sincere agnostics.

[45] On this transition see Cardinal Jean Daniélou's 1964 study of *The Theology of Jewish-Christianity*.

extended this patristic theology to include Greek natural science in the somewhat more developed form given it by Islamic thinkers.[46] Has the theology of the period since Vatican II been as successful in assimilating modern science?

We regret to say that recent Catholic theologians, while they have done much to assimilate the findings of modern history, psychology, and the social sciences to theology, they are still largely content to leave the physical and biological sciences go their own way. The Jesuit paleontologist Teilhard de Chardin, particularly perhaps in the book he published in what turned out to be his last year of life, 1955, *Le Phénomène Humain*,[47] had prophetically raised the question of how theology is to face the discovery that the universe as we know it today — both as a whole and in its specifically human biological and cultural dimensions — is a product of a whole complex of evolutionary processes;[48] but his own efforts at advancing a

---

[46] See "The contribution of Islam to Philosophy in the Latin Age", in Deely 2001: 186–193.

[47] "This work may be summed up", Teilhard begins it by telling us (English ed., p. 31), "as an attempt to see and to show what happens to man, and what conclusions are forced upon us, when he is placed fairly and squarely within the framework of phenomenon and appearance". The English translation of this work by Bernard Wall, *The Phenomenon of Man*, appeared in 1959. See Deely 1966a: "The Vision of Man in Teilhard de Chardin".

[48] Nor is it only theology, but traditional philosophy too that has to come to terms with the cosmos as we have learned it to be — not an unchanging heavens centered on a stationary earth, but an evolutionary whole which does not even have a center that we can determine in our present awareness. The resistance to this change in "background cosmological image", as it were, for *all* thought, scientific, philosophical, or theological, is a matter of record, well documented, for example, in Stimson's little book of 1972, *The Gradual Acceptance of the Copernican Theory of the Universe*. Nor is the early modern accumulation of abuses on the part of authorities civil and religious, encouraged by the late Scholastics to take actions of repression and thought-control, a record of which we can be proud: see "The crash and burn of Scholasticism, c.1600–1650" (Deely 2010b: 381–384). Philosophy, too, insofar as it has continuity with the medieval period, has a "coming-to-terms" to reckon with in this matter of the discoveries of modern science, as the 1963 work of Raymond Nogar, *The Wisdom of Evolution*,

solution have met little success, being generally either stone-walled or actively resisted. In the United States, the Benedictine Stanley L. Jaki (1924–2009) has provided an honorable exception to this silence, but most Catholic theologians have ignored also his valuable work[49] — though note must certainly be taken of Pope Benedict XVI's Homily in the Cathedral of Aosta, Italy, on 24 July 2009, where he said:[50]

> The role of the priesthood is to consecrate the world so that it may become a living host, a liturgy: so that the liturgy may not be something alongside the reality of the world, but that the world itself shall become a living host, a liturgy. This is also the great vision of Teilhard de Chardin: in the end we shall achieve a true cosmic liturgy, where the cosmos becomes a living host.

Within science itself there are different fields of specialization, often using different language and methods; yet scientists are usually not content with theories in their own field that cannot be integrated with the basic principles of physics. If scientists cannot see a true integration of the worldview of religion with the world that their scientific research reveals to them, their dialogue with religion will never be free of suspicion. The more integrative approach to the relation of modern science and theology, that we are attempting in

---

meticulously outlined. (See also Deely and Nogar 1973: *The Problem of Evolution*; Deely 1965/66: "Evolution: Concept and Content", and 1969: "The Philosophical Dimensions of the Origin of Species"; and Ashley 1973: "Change and Process".)

[49] We might note in particular Jaki's 1989 book, *God and the Cosmologists*; but, as is noted in the *Wikipedia* <http://en.wikipedia.org/wiki/Stanley_Jaki>, Jaki "authored more than two dozen books on the relation between modern science and orthodox Christianity".

[50] Celebration of Vespers with the Faithful of Aosta, Homily of Pope Benedict XVI <http://www.vatican.va/holy_father/benedict_xvi/homilies/2009/documents/hf_ben-xvi_hom_20090724_vespri-aosta_en.html>. See the *NCR* article by Allen 2009, "Pope cites Teilhardian vision of the cosmos as a 'living host'," with readers' comments.

this work, does not exclude dialogue, nor deny the relative independence of science and theology, nor even temporary conflicts between the two, but is based on the conviction that only if theology is willing to profit from the work of scientists is it likely that scientists might also be willing to profit from the work of theologians.

Theology, St. Thomas Aquinas argues[51] is a true "science" (*scientia*) in the sense of critical knowledge that arrives at some solid results. He argues also that this theological science is primarily theoretical (that is, contemplative of "how things are"), but by reason of this very fact it is also eminently practical in that it leads us to our true fulfillment, the perfect contemplation of God while living a life based on faith as comprehending all around us. God is absolutely One, but this unity is communicated from the Father to the Son in the Holy Spirit, a communication manifested to us in the "economy" or holy history of the Biblical narrative summarized in the Creed. Since the Church's Creed is Trinitarian, theology's principles must be derived from that summation of the Church's faith tradition.[52]

Therefore, it seems fitting to consider how our reason in the form of progressive scientific knowledge can enrich theology by reflecting on how it manifests the Father, then the Son, and then their unity in the Holy Spirit as Creator, Redeemer, or Sanctifier, Love, Compassion, Grace, or by other such positive terms.[53] We will show, moreover, that the traditional, biblical terms viewed in the light of the modern biology of sexual differentiation are a powerful refutation of sexism. Furthermore the Blessed Virgin Mary, "chief member" of the Church in the 1964 declaration of Vatican II,[54] as she is Mother of God,

---

[51] Aquinas 1266, *Summa theologiae* I, q. 1. art. 4.

[52] On this see Gilles Emery 2007, *The Trinitarian Theology of Saint Thomas Aquinas*.

[53] Cf. John Cobb 1975, *Christ in a Pluralistic Age*.

[54] *Lumen Gentium*, Chap. 8, no. 53: "Wherefore she [Mary] is hailed as a pre-eminent and singular member of the Church and as its type and excel-

is also Mother of the Creation; and the creation, the moral law, the People of God personified, is often spoken of as Wisdom in the Scriptures. Thus scientific knowledge of creation must enrich Mariology.[55]

This book is a writing of theology, not of experimental science, but a work of theology using the common views of current scientists. Fr. Ashley in particular, as a theologian and a member of the Dominicans or Order of Preachers, writes in the tradition of St. Thomas Aquinas, whose views both as to reason and to faith have so often been recommended by the Popes. The two greatest medieval theologians, the Franciscan St. Bonaventure, and the Dominican St. Thomas Aquinas, both[56] showed how the Trinity is reflected in the creation of

---

lent exemplar in faith and charity. The Catholic Church, taught by the Holy Spirit, honors her with filial affection and piety as a most beloved mother". Mary "is endowed with the high office and dignity of being the Mother of the Son of God, by which account she is also the beloved daughter of the Father and the temple of the Holy Spirit. Because of this gift of sublime grace she far surpasses all creatures, both in heaven and on earth. At the same time, however, because she belongs to the offspring of Adam she is one with all those who are to be saved." She is "the mother of the members of Christ ... having cooperated by charity that the faithful might be born in the Church, who are members of that Head". Cf. Pope Pius XII's 1947 constitution "Provida Mater", and his "Annus Sacer" allocution of 1950.

[55] See Ashley 2011a, *A Marian Ecclesiology*, which aims to answer those who hold that Jesus intended to start a "movement" but not an "organized church".

[56] For St. Bonaventure's view see Diagram 1, p. 131, of Sister Paula Jean Miller's 1996 work, *Marriage: The Sacrament of Divine-Human Communion: A Commentary on St. Bonaventure's Breviloquium.* For St. Thomas's view see his *Summa theologiae* I, q. 45, art. 7: "Now the processions of the divine Persons are referred to the acts of intellect and will .... For the Son proceeds as the word of the intellect; and the Holy Ghost proceeds as love of the will. Therefore in rational creatures, possessing intellect and will, there is found the representation of the Trinity by way of image, inasmuch as there is found in them the word conceived, and the love proceeding. But in all creatures there is found the traces of the Trinity, inasmuch as in every creature are found some things which are necessarily reduced to the divine Persons as to their cause. For every creature subsists in its own being, and has a form, whereby it is determined to a species, and has relation to something else.

the universe and humankind, in the unity of community and mutual relationship as well as in the unique depths of each human person. May they assist and guide us in what follows!

Therefore as it is a created substance, it represents the cause and principle; and so in that manner it shows the Person of the Father, Who is the 'principle from no principle'. According as it has a form and species, it represents the Word as the form of the thing made by art is from the conception of the craftsman. According as it has relation of order, it represents the Holy Spirit, inasmuch as He is love, because the order of the effect to something else is from the will of the Creator. And therefore Augustine says (*De Trinitate* vi 10) that the trace of the Trinity is found in every creature, according 'as it is one individual', and according 'as it is formed by a species', and according as it 'has a certain relation of order'. And to these also are reduced those three, 'number', 'weight', and 'measure', mentioned in the Book of Wisdom (9:21). For 'measure' refers to the substance of the thing limited by its principles, 'number' refers to the species, 'weight' refers to the order. And to these three are reduced the other three mentioned by Augustine (*De Natura Boni* iii), 'mode', 'species', and 'order', and also those he mentions: 'that which exists; whereby it is distinguished; whereby it agrees'. For a thing exists by its substance, is distinct by its form, and agrees by its order. Other similar expressions may be easily reduced to the above."

Chapter 2

THE IMMENSITY, VARIETY, AND DYNAMISM OF THE UNIVERSE REVEAL GOD THE FATHER[1]

How Big is Our Universe?

In this first part of our exploration of how modern science can enrich Christian theology, we will aim to confine our analysis to the universe as we observe it today in our earthly present. We will pass over the theories of its history, taking up only the prevailing scientific consensus today. Its cosmic and biological evolution will be taken up later in Chapter 4. Since current scientific theory is strongly evolutionary, to stick with the present in this way may seem very odd. The reason we have chosen to do so is because we can only know the past and future through the present; and by the Principle of Research, mentioned in Chapter 1, it is always best to be careful to proceed from what is better known to the less known.

The power of God, attributed to the Father as the Principle (the "First Person") of the Trinity from whom the Son and Holy Spirit proceed is manifested in three necessarily related ways: (a) by the magnitude of the whole creation; (b) by the variety of the substances that compose it; and (c) by the dynamism of their interactions, that is, by the ways in which

---

[1] The authors wish to express particular thanks to Professor James Clarage of the Physics Department of the University of St Thomas, Houston, for his careful reading of and comments and suggestions for the final text of this chapter.

creatures share in the divine power. The first of these, the magnitude of the universe, does not consist only in its quantitative size, although that is included, but more importantly in the *variety and complexity* of the substances, material and immaterial, that compose it.

Of any material thing, the first property that is evident to us in the continuing process of material change is its *quantity*. How large is this mass of matter, dark and luminous, that is our universe? Some believe that the part of the universe today visible to astronomers has a radius of 13.7 billion light years and seems to be expanding at an accelerating rate. The portion of the material universe now visible, however, is much smaller, perhaps only $1/50^{th}$ of its real totality, because we can observe only that portion whose light has by now reached us.[2]

Some suppose that this expanding universe very early in its history also very briefly "inflated" very rapidly,[3] and may even have become infinite in size; but an *actual infinite* in this sense is unobservable, for our actual knowledge of remote parts of the universe depends upon those remote parts having effects upon the parts that we can actually observe directly, that is to say, upon the finite region in which scientific research is based; and at least this region of the universe visible to us is a relatively balanced system of bodies interacting with each other, itself finite, but of which we may say that it is "infinite" only in the sense in which one thing after another is always coming into existence within an interacting system expanding as a whole but without there being any prejacent "limit" to be reached

---

[2] Britt 2004 estimates width of cosmos at 156 billion light years. But cf. the Wiki "observable universe" article, the "misconceptions section": <http://en.wikipedia.org/wiki/Observable_universe#Misconceptions>.

[3] "Cosmic inflation", "cosmological inflation", or just "inflation", is supposed to have been an exponential expansion of the universe 10^{-36} seconds after the Big Bang, driven by a negative-pressure vacuum energy density. On this see Alan H. Guth 1981 and 1997: "Inflationary Universe"; also <http://en.wikipedia.org/wiki/Cosmic_inflation>.

outside of itself. Hence the surprising analytical conclusion reached by Aristotle still holds:[4]

> The infinite turns out to be the contrary of what it is said to be. It is not what has nothing outside it that is infinite, but what always has something outside it. ... [Hence] Our definition then is as follows: *A quantity is infinite if it is such that we can always take a part outside what has been already taken.*

Today it is believed that about 95% of the massive material of the universe is "dark matter" that gives off no light, and hence can be observed only by its gravitational attraction on visible matter. This dark matter is, however, probably in more or less random motion. Because it has mass it tends to consolidate, but this tendency is opposed by another mysterious force called "dark energy", which is driving all massive objects ever further apart at an accelerating rate.[5]

The other 10% of the matter in the universe that can be detected by the light it emits or reflects is about 90% hydrogen and 10% helium gas, He, along with a trace of lithium, Li. Within the cores of red giant stars, such as our own sun will some day

---

[4] Aristotle i.353/347BC: *Physics* III ch. 6, 206b35–207a1, and 207a7–8. See further ch. 6, 206a27–29: "For generally the infinite has this mode of existence: one thing is always being taken after another, and each thing that is taken is always finite, but always different." Hence (206a18, 21, 23) "... the infinite has a potential existence. ... There will not be an actual infinite. ... we say that the infinite 'is' in the sense in which ... one thing after another is always coming into existence" within an ever-expanding system.

Aristotle treats of the question of cosmic infinity also in his c.355BC treatise *On the Heavens* (*De Caelo*), Book I, Chaps. 5–7. His treatment of the problem of the infinite in his *Physics* occupies all eight of the chapters of Book III, particularly (203b22–24) "the difficulty that is felt by everybody — not only number but also mathematical magnitudes and what is outside the [observable] heaven are supposed to be infinite because they never give out in our *thought*." Hence (ch. 8, 207b27) "Our account does not rob the mathematicians of their science".

[5] On current theories, see Sato 2009: "Is Dark Matter & Dark Energy the Same Thing?".

become, the synthesis of atoms heavier than hydrogen, helium, and lithium is in process. These heavier atoms constitute the remaining approximately 2% of the visible matter of the universe. For the possibility of life on earth the most important of these minute bits of matter within stars are carbon atoms, each forged by the strong nuclear force from three helium atoms. Such forging can occur, moreover, only in very narrow energy conditions or the nuclei will again decompose. When a red giant star finally explodes as a planetary nebula these heavier atoms are scattered into interstellar space. Elements heavier than iron are synthesized from the extreme conditions at the death of more massive stars, which explode as a supernova.

Water, H_2O, contains elements from both the beginning of time and the present time: hydrogen from the Big Bang, and oxygen from the ongoing stellar processes mentioned above. The water molecule, essential for life, has been detected within the great clouds of cosmic dust whose grains are less than 0.1 mm in size that fills all of interstellar space. The stars of various sizes and in various states of formation and extinction, some in great clusters, are gathered in immense rotating clouds or galaxies of various shapes, probably centering on a "black hole" of density so extreme that it no longer emits light. These galaxies are also collected in numerous, elongated super-galaxies in filaments and walls between which are vast deserts of featureless matter and "empty" fields of electromagnetic and gravitational forces. Within the galaxies the stars often cluster, sometimes by thousands, and are surrounded (like our sun, a typical star) by moving materials — planets, asteroids, comets, etc., — held together by gravitational "attraction".[6]

While water exists in some form in many places throughout the universe, remarkably, it covers some 70% of our earth's surface, while on the other planets in our system it is present mainly

---

[6] "Attraction" names the apparent force gravity exerts, but, remarkably, the real nature of gravity remains elusive to our science so far! See esp. pp. 39–43 following in the discussion of Relativity Theory.

as ice or a gas.[7] The human body is by weight about 75% water; the blood 95%, lean muscle 75%, bone, 22%, body fat 14%, with some 10% more water in adult men than adult women. No wonder then that in the liturgies of most cultures washing in water is a ritual of purification, and in Christianity it is the fundamental Sacrament of Baptism, commanded by the Risen Christ to the Apostles before his Ascension. (Mt 28:18–20):

> All power in heaven and on earth has been given to me. Go, therefore, and make disciples of all nations, baptizing them ["in water" — Acts 8:36, 37] in the name of the Father, and of the Son, and of the Holy Spirit, teaching them to observe all that I have commanded you. And behold, I am with you always, until the end of the age.

The symbolism of this sacramental rite is universal. A child comes into existence in the water within its mother's womb, and death reduces the body toward dust which washes away to the ocean depths into which all rivers flow. So the Sacrament of Baptism washes away all sin and infuses into the baptized the new life of grace. Thus the Christian sacramental system is in harmony with and reflective of the laws of nature, even though it is of a far higher order.

As indicated in Chapter 1, two great mathematical constructs are used by physicists today to explain this vast universe as an interacting system, namely, the Special and General Theory of Relativity at the macro-level, and the Standard Model of Quantum Physics at the micro-level.

Relativity Theory

Relating primarily to the macro-level of material reality is the Special and General Theory of Relativity, developed by

---

[7] See "Water in the Universe" <http://www.ozh2o.com/h2universe. html>. Yet bear in mind the notable exception of Jupiter's moon, Europa, which may harbor a water ocean kept liquid by gravitational "heating" caused by tidal forces.

Albert Einstein in 1905–1915 and now generally accepted as supported by extensive evidence. This theory is based on two major factual claims. The *first* factual claim is that the speed of light in interstellar space is constant, about 186,282.397 miles per second, or 670,616,629.384 miles per hour. Thus, in a year light travels a "light-year" of 5,865,696,000,000 miles, and makes possible our observation and establishment of the positions and motions of distant bodies. *Second*, since we have no way of determining whether any body in the universe is absolutely at rest, all our scientific measurements of the motion of bodies depends not simply on three spatial dimensions but on a fourth factor, namely, the time that the light they emit takes to reach us. Hence the measurement of time must be treated mathematically as a *fourth dimension* of physical reality in its knowability to us. In this way, the laws of motion can be considered as the same for all observers in relation to their own frame of reference, *as if* they were at rest.

From this relativistic view of the measurement of motion certain surprising results follow. *First*, the paths of bodies moving in this space-time seem to converge, and hence space is said to be "curved". Gravitational "attraction" with its acceleration of velocity, famously proved by Galileo and fundamental to Newtonian astronomy, is now attributed simply to this "curvature" of space-time as it affects the path of a moving body in relation to another such body. *Second*, because of the constant speed of light, as bodies approach that speed they contract in the direction of motion (the Lorentz Contraction)[8] and increase in mass (resistance to that motion). *Third*, since this acceleration requires increasing energy, the measurements of "matter" (defined by the equivalence of gravity and inertia or resistance to motion) and of energy are convertible ($E=mc^2$).[9] Hence, in

---

[8] See "Length Contraction" in <http://en.wikipedia.org/wiki/Length_contraction>.

[9] See "Mass-Energy Equivalence" in <http://en.wikipedia.org/wiki/Mass-energy_equivalence>.

this relativistic picture, the laws of physics hold for any observer — but the real state of affairs at the macro-level of bodies as we observe them remains subject to various interpretations.

This crucial reservation, of course, held true also for the ancient astronomers who, as Ptolemy admitted, were simply (following Plato's original recommendation) "saving the appearances" by mathematical constructions consistent with observed measurements. This was the qualification that the Inquisitors demanded that Galileo make respecting his affirmation of Copernicus' heliocentric theory; but later, when Robert Bellarmine (†1621) as Chief Inquisitor was dead, and one of the Italian Cardinals, namely, Maffeo Barberini (1588–1644) — who had shown public support for Galileo at the time of the latter's visit to Rome in 1616 — had been elected (6 August 1623) as Pope Urban VIII, Galileo deemed it time to ignore the secret oath that Bellarmine had privately but officially extracted from him in writing, an "agreement" for Galileo to cease proposing that a Copernican view (i.e., a view that the earth moved relative to the sun rather than *e converso*) is more than a hypothesis.[10]

---

[10] For the often misrepresented facts of this case, see Fantoli 2003: *Galileo: For Copernicanism and For the Church*. Indeed, this iconic case of the early modern era bears on our very purpose for writing the present book. "As they returned to their homes after the Congregation's session", Fantoli notes (1996: 427), "for the cardinals and officials of the Holy Office the 'Galileo affair' was by now a closed chapter. They had no suspicion that the true 'Galileo affair' was instead beginning right on that very day 22 June 1633 and that their names would pass to posterity not only as judges of the tribunal of the Holy Office but also and above all as the accused, destined to be called innumerable times in the centuries to come before the much more severe tribunal of history." Thus repeated arguments have been made on behalf of the Vatican Inquisitors of that day to the effect that Galileo was guilty at the time of making assertions he could not prove — as if the accusers of Galileo by contrast were resting their case on quite provable assertions and solid theological grounds. Yet how does it improve or vindicate or render palatable the false certainty of the Inquisitors concerning what was 'of Faith', to note that not until 1728 did Bradley record an aberration of starlight that gave the first direct evidence for the revolution of the earth around the sun? Or to note that not until 1818 was Bessel able so to correct and organize Bradley's data as to factually establish

That the macro-level of bodies as observable by us, by the way, remains subject to various interpretations remained true, but in a different way, for Newtonian astronomy, which

---

parallax, the apparent motion of the stars theoretically required by motion of the earth?

The historical record of scientific advance subsequent to the trial of Galileo hardly changes the circumstance that Copernicus' work had earlier been declared "heretical" — i.e., incompatible with Christian faith — on a question that could not legitimately be made subject to such a judgment. (On the Scriptural texts appealed to at the time, see Deely 2001: 494 note 11; but see also note 10, last paragraph.)

The larger point of the Galileo case, surely, is much more *prospective* than retrospective — more postmodern than modern. The larger point is not some secret workings of the past, but that there are limits to religious authority and to the understanding of all doctrines, theological as well as philosophical and/or scientific. The problem of finding out more clearly where and what those limits are with respect to science in particular has been the principal story of modern times as the epoch of civilization in which science established its ideoscopic perspective as distinct from the comparatively cenoscopic perspectives of philosophy and theology alike.

Thus the "postmodern" view is that modern times saw the establishment of ideoscopic science as distinct from cenoscopic science, as medieval times saw the establishment of theology as distinct from philosophy, and ancient times saw the realization of philosophy as distinct from mythology: overall, the record of a long painful growth from childhood toward a maturity that will come, no doubt, yet largely still eludes the species wishfully styled 'sapiens'. Indeed, the Enlightenment looks today like the adolescence of intellectual culture in coming to terms with modern science! See "Beyond the Latin Umwelt: Science comes of Age", Chap. 11 in Deely 2001: 487–510, particularly "The Condemnation (21 June 1633) of Galileo Galilei", 493–499. "The judges of the Holy Office erred so gravely" in their treatment of the Galileo affair, say Jacques Maritain (in his discussion of 1973: "The Error of the Holy Office", in *On the Church of Christ*, pp. 209–210, emphasis supplied), as in their earlier condemnation as "heretical" the doctrine of Copernicus, "through an error of principle still more dangerous because of its general bearing, [namely,] *they held the science of phenomena in its own development to be subject to theology, and to a literal interpretation of Scripture* against which St. Augustine and St. Thomas has forewarned us" — to which he adds the assessment of Cardinal Journet 1955: *The Church of the Word Incarnate*, p. 232): the clerics of Galileo's time "lacked the courage needed to detach the question of Scripture at once from the dispute over the geocentric issue".

explained planetary motion by "action at a distance" in "absolute space". More cautious on the point than Galileo had been, Newton simply attributed the overall situation to the action of God.[11] Similarly today, Relativity Theory "saves the appearances", but gives us only a *perspectival* and mathematically constructed view of the universe in its total reality.

Quantum Theory

The second great mathematical construct developed in the 20th century that achieved a similar "saving of the appearances", but now at the micro-level of physical reality, is the standard theory (or "model") of quantum physics.[12] Already at the very opening of the 19th century, Thomas Young (1773–1829) in 1801 had established experiments which implied the wave-like nature of light.[13] Later experiments by James Clerk Maxwell (1831–1879) confirmed Young's work, but Maxwell's later mathematical predictions that light might have an electric + magnetic composition also showed that this E + M disturbance carried momentum like a particle — which is odd for a wave. Still, this scientific work of the 19th century served to show that light is not a stream of particles, as Newton had supposed, but rather a wave phenomenon seeming to travel in a mechanical aether. This wave-like electromagnetism differs from gravitation, and is the second fundamental force in the

---

[11] See Lindberg and Numbers 2003: *When Science and Christianity Meet,* 2003), pp. 82–83.

[12] The Quantum Theory was more of a collective work than had been Einstein's achievement of relativity theory, involving the work of such physicists as Max Planck (1858–1957), Einstein himself, Niels Bohr (1881–1962), Max Born (1881–1970), Erwin Schrödinger (1887–1961), Louis de Broglie (1892–1987), Werner Heisenberg (1901–1976), Paul Dirac (1902–1984) and John von Neumann (1903–1957). The theory was refined in the second half of the 20th century by Richard. P. Feynman, (1918–1988), Freeman Dyson, (1923–), Abdus Salam (1926–1996), Stephen Weinberg (1933–), among others.

[13] See Parry-Hill and Davidson 2006: "Thomas Young's Double Slit Experiment".

universe as understood by science today. While gravitation is "attractive" (or "collapsive"! — see note 6 above), electromagnetism, although it too produces attraction between particles oppositely charged electrically, also creates repulsion between particles of the same charge. (A fuller account of the complicated classification of these particles, as they have been progressively discovered by breaking up atoms through raising the energies of the particles of which they are composed, we will take up in the section following, pp. 47–53).

These minute "particles" of which all matter is composed, except perhaps the so-called "vacuum", are measured by what is now considered the ultimate natural unit of length, to wit, the minute "Planck length" of 1.6×10^{-35} meters. This unit is itself defined by three fundamental physical constants: the speed of light, Planck's constant, and the gravitational constant. The two fundamental types of particles, both so small that they are treated mathematically simply as points in space, are *quarks* (quantified in protons[14]) and *leptons* (typified by *electrons*). A mathematical formula that seems to describe not a linear but a wave motion, called "Schrödinger's Equation", has been shown to predict very accurately the motion of such particles at the micro-level, yet it remains that at the macro-level particles and waves have very different kinds of motions!

The realistic interpretation to be given this puzzling "wave-particle duality" remains controversial. Bohr, Born, and Heisenberg, influenced by Kantian idealism, discovered that (according to the probabilities set by Schrödinger's equation) a moving particle can be said to have a determined position only when actually observed, thus leaving its presence *during* its motion vague, uncertain. Thus, in our observed universe, it seems that elementary particles in motion have paths that can only be predicted by *a range of probabilities* that is best represented mathematically by a wave.

---

[14] Owing to the swarming motion of the fundamental point-like quarks, the proton achieves a "size" of about 10^{-18} meters.

It is clear, therefore, that *both* our macro and micro views of our universe have been immensely deepened by modern science, but only at the expense of relying on mathematical constructs that "save the appearances", yet whose realistic interpretation remains obscure and much debated. Moreover, as we noted in Chapter 1, to date these two great theories of Relativity and Quantum physics have not been unified by a "Final Theory". The most promising candidate for such a theory so far is String Theory. String Theory has many versions, none of which picture elementary particles as zero-dimensional points, but rather as one-dimensional objects set in ten or eleven space-time dimensions which are, however, "curled up", so as not to be observable.[15] In this construction, all the fundamental forces are reduced to one. Already the so-called weak force, responsible for the instability of radioactive elements, has been joined to electromagnetism as the *electro*weak force; but the strong nuclear force has not. As we also noted in Chapter 1, this String Theory, on which physicists have now been working for almost half a century, though promising, still has not been verified because of the immense energies required to test it; and it may prove a dead end.[16]

Jews, Muslims, and Christians do not accept any of the various forms of monism that favor a single spiritual Absolute Reality, whether they are scientistic materialism or whether they are spiritual pantheism, including (cf. Deely 2001: 288–289) the "panentheism" of Buddhism. Instead, these monotheistic religions see the vastness of the universe as a reality manifesting by *analogy* the transcendence of the Creator in relation to the Creation. He has created it in entire freedom for its own sake, and not to satisfy any need on his own part. Thus, in

---

[15] See Groleau 2003: "Imagining Other Dimensions", at <http://www.pbs.org/wgbh/nova/elegant/dimensions.html>.

[16] See "String Theory" in the *Wikipedia* <http://en.wikipedia.org/wiki/String_theory>. A *severe* criticism of the String Theory hypothesis is offered by Woit 2002: "Is String Theory Even Wrong?"

the Bible, written by Jews whose knowledge of the size of the universe was quite primitive, amazement at God's creation is nonetheless expressed again and again:

> Thus the heavens and the earth were completed in all their vast array [Gn 2:1];

> Mighty Lord ... You rule the raging sea; you still its swelling waves [Ps 89:10].

God says to Job (38:16–20):

> Have you entered into the sources of the sea, or walked about in the depths of the abyss? Have the gates of death been shown to you, or have you seen the gates of darkness? Have you comprehended the breadth of the earth? Tell me, if you know it all: Which is the way to the dwelling place of light, and where is the abode of darkness that you may take them to their boundaries and set them on their homeward paths?

Isaiah and the other prophets cry out (Is 49:13):

> Sing out, O heavens, and rejoice, O earth, break forth into song, you mountains.

This analogy of the material size of the universe to God's infinite power, while it is in a way crude, is fundamental to our understanding of who and what God is, and our theology must expand to meet it. Later we will see that it is probable that if the universe had not expanded to such great dimensions life could not have emerged within it.

The first conclusion of this chapter, therefore, is that, although the Church has always taught that the Power of God is infinite, modern science, by proving the quantitative vastness of the universe as a dynamic system, has made it possible for us to praise God as Power in a concrete and detailed way that the people of former times could only and in the comparatively *vaguest* terms imagine.

How Varied is Our Universe?

We have listed some of the simplest entities that have, in some sense, independent existence. The Bible names various kinds of inanimate substances. Of special symbolic importance was the breastplate of the high priest Aaron that had four rows of jewels, three in each row for the twelve tribes of Israel, that were probably ruby, topaz, carbuncle, emerald, sapphire, diamond, jacinth, agate, amethyst, beryl, onyx, and jasper (Exodus 39:10–13); but there is little agreement among translators. The prophet in *Ezekiel* 28:13 uses this breastplate and its jewels to describe human perfection as it supposedly was in the Garden of Eden, and the *Book of Revelation* 21:18–20 uses it to describe the walls of the Heavenly City of Jerusalem and its gates of pearl. Other inanimate substances of biblical prominence are air, alabaster, amber, antimony, asphalt, brimstone, bronze, carnelian, cedar-wood, chalk, chrysoberyl, chrysolite, chrysophrase, cinnabar, clay, coal, copper, coral, earth, fire, flint, garnet, glass, hyacinth, incense, iron, ivory, lapis lazuli, lead, ligure, lime, lachite, light, marble, mica, moon-stone, myrrh, niter, oil, opal (bdellium), perfume, pitch, rock, salt, slime, smoke, stone, tin, turquoise, vinegar, water, wine, wood, and zircon.

As for living things (in which the Bible takes more interest), long before the development of science humans marveled at their variety, named them, and tried to understand their different natures, as we see in the cave-paintings of early man. The books of the Bible are marked by this same vivid interest:

God said: "See, I give you every seed-bearing plant all over the earth and every tree that has seed-bearing fruit on it to be your food; and to all the animals of the land, all the birds of the air, and all the living creatures that crawl on the ground, I give all the green plants for food". [Gn 1:30]

So the LORD God formed out of the ground various wild animals and various birds of the air, and he brought them to the man to see what he would call them; whatever the man called each of them would be its name [Gn 2: 19].

We now know that this "bringing to be" was also — at the level of secondary causes respecting the "creation from nothing" whereby the First Cause maintains the very existence of the complex of interacting finite beings that comprise the creation — the result of *evolutionary processes* in which environmental changes eliminate some species and favor new ones, some places rising to intelligent life and other places not. Can there be higher forms of intelligent life that are, like ours, dependent on bodies to function? Again perhaps. Efforts by astronomers — for example, the SETI ("Search for Extraterrestrial Intelligence") project — to communicate with such intelligences have as yet yielded no firmly positive results; but the fifty some years of this project are (to say the least) rather brief in time and extent within the scale of the universe as today we have come to know it!

The Biblical writers — to say the least — were not at all aware of what modern science has revealed of the scale of space and time; even so, they did greatly appreciate the biological diversity of their Near East surrounding. Merely to list the biblical names of living things other than human persons gives some idea of its authors' intense interest in them, although the translation of these names is often uncertain. Over one hundred kinds of plants, not always easy to identify, are mentioned in the Bible, including: such *trees* as the acacia, almond, apple, ash, balsam, bay-tree, box-tree, cedar, chestnut and other nut-trees, cinnamon tree, cypress, date-tree, elm, fig-tree, fir-tree, gum-tree, hazel-tree, hemlock, juniper, mulberry, myrtle, oak, oil-tree, olives-tree, palm, pine, plane-tree, poplar, sandalwood, shittah-tree, storax-tree, sycamore, terebinth, thorn-tree, thyine-tree, willows; *cultivated crops* such as barley, beans, corn, flax, grass, millet, rye, sweet-cane; *bushes* such as brambles, brier, bulrush, cane, nettles, reeds, rushes, thistles; *vegetables* such as beans, cucumbers, garlic, gourds, leeks, lentils, onions; *fruits* such as apples, mandrakes, melons, grapes, pomegranate; *flowers* such as the rose, hyacinth, and the lily; *herbs and spices* such as cassia, coriander, mustard, rue, fitches; *perfumes* such as balm of Gilead, galbanum, myrrh, and spikenard; *dyes* such as saffron; and *weeds* such as cockle, nettles, and thorns, and even mildew and rot.

Among over one hundred species of animals that the Bible mentions are marine coral, fish, snails, a great variety of worms and snakes, frogs, lizards, chameleons and crocodiles; many kinds of insects such as ant, bee, beetle, cricket, cochineal, flea, fly, grasshopper, locust, louse, mosquito, moth, spider, wasp; snakes of many kinds, lizards, leeches, frogs, chameleon; many kinds of birds: bitterns, buzzards, cormorants, cranes, cuckoos, doves, eagles, hawks, herons, ibises, lapwings, nighthawks, kites, owls, ostriches, partridges, peacocks, pelicans, quail, ravens, seagulls, sparrows, storks, swallows, swans, vultures; small animals like rabbits, hyraxes, hedgehogs, badgers, moles, porcupines, weasels, in contrast to mighty elephants, hippopotamuses, and rhinoceroses, along with the more familiar antelope, deer, cattle, gazelles, goats, sheep, asses, horses, mules; swine and camels, the odd giraffe; cats, lions, leopards, dog, wolves, jackals, foxes, hyenas, the timid rabbit, conies, the flying bats, the whales and seals that have returned to the ocean; and finally our close imitators the monkeys, baboons, and apes. While it is evident that much of this biblical knowledge arose from practical concerns, it also exhibits wonder and curiosity. As Jesus said, in the Sermon on the Mount (Mt 6:26–30):

> Look at the birds in the sky; they do not sow or reap, they gather nothing into barns, yet your heavenly Father feeds them. Are not you more important than they? Can any of you by worrying add a single moment to your life-span? Why are you anxious about clothes? Learn from the way the wild flowers grow. They do not work or spin. But I tell you that not even Solomon in all his splendor was

clothed like one of them. If God so clothes the grass of the field, which grows today and is thrown into the oven tomorrow, will he not much more provide for you, O you of little faith?

Here we must ask what modern science has added, beyond our vastly extended knowledge of biodiversity, to the further physical and chemical diversity involved within and supportive of the biological realm? As already noted, in current science there are four fundamental forces: gravity, electromagnetism, and the weak and strong forces that bind the nuclei of atoms. According to the Standard Model[17] that Steven Weinberg, as we cited him in Chapter 1,[18] so praises for its logical consistency and extensive empirical verification, these entities are classified as either simple (without any internal structure) or complex. The simple ones are either:

1) *fermions* that constitute massive matter and have ½ spin. By "spin" is meant a *directional* property of particles that helps to specify them but whose exact nature, like much about these particles, remains obscure. Fermions are sub-divided into

   (1.1) *quarks* that react with each other by what is called the "color" force with six "flavors" (up-down, strange-charm, bottom-top) each of which has an anti-quark.

---

[17] Modifications to The Standard Model are implied in the development of what is called a "Supersymmetry", an attempt to unify the four fundamental forces principally by the still-experimentally-unconfirmed String Theory (note 16 preceding). This theory suggests the existence of many other hypothetical particles, such as the neutralino (spin -½) (a leading candidate for dark matter); the photino (spin -½), superpartner to the photon; the gravitino (spin 1/3), superpartner of the graviton; sleptons and squarks (spin 0), superpartners of the fermions. The hypothetical graviton (spin -2) is needed to mediate gravity, while the more speculative magnetic monopoles would have non-zero magnetic charge. It is premature to comment on the status or implications of this theoretical proposal, so we simply make note of it here "for future reference".

[18] See, in Chapter 1, pp. 13–28, esp. p. 14.

(1.2) *leptons* also with six "flavors", each with an anti-lepton and which are:

(1.2.1) charged positively or negatively and of six kinds: *electron/positron*, negative/positive *muon*, and negative/positive *tau*.

(1.2.2) uncharged *neutrinos* also with their anti-neutrinos and in six varieties.

2) *bosons*, with integer or zero spin, that transmit forces between massive bodies,

(2.1) *gauge bosons*, *photon* (light particle) mediating electromagnetism and *gluon* (strong force binding quarks), both massless.

(2.2) *W bosons*$^\pm$ mediating weak nuclear force and *Z bosons*^{0} mediating strong nuclear force, both massive.

(2.3) hypothetical *Higgs boson*, massless, spin 0, conferring mass on other elementary particles.

These elementary particles can form composite particles, classified either as strongly interacting composites that are either:

1. Hadrons

1.1. *Baryons* that are massive (fermions) containing three quarks or antiquarks. Examples are the positive *proton* and the neutral *neutron* that form the nucleus of atoms.

1.2. *Mesons* that are massive (bosons) with integral spin containing a quark and an anti-quark of which there are many possible combinations historically referred to as "the particle zoo" before the underlying quark model was discovered.

2.2. Atomic nuclei.

2.3. Atoms.

2.4. Molecules.

2.5. Macromolecules, e.g., the hundreds of thousands of unique polymeric combinations of simple molecules synthesized by cellular life processes. Well-known examples are DNA, RNA, and proteins.

These particles, other than those of dark matter that must be extremely inert, are observed independently only when the atoms in which they ordinarily exist are broken up by high energy collisions in an accelerator. Quarks, however, have never been observed in isolation, but only within atomic nuclei, mesons, or high temperature quark-gluon plasmas. As already indicated, in the 5% of the universe's matter that is not dark matter, atomic hydrogen gas is about 75%, commonly in the form of diatomic molecules H_2, which is simply two hydrogen atoms bonded together, while the rest consists in about 24% helium, 1% deuterium (an isotope of hydrogen), and traces of lithium and beryllium, but no heavier elements.

As already noted, the nature of the dark matter is still unknown but it may be made up of the humorously named WIMPS (Weakly Interacting Massive Particles) or MACHOS (Massive Compact Halo Objects, dark objects with inertial mass, such as dust or even brown dwarf stars) in the haloes surrounding galaxies, as well as axions (hypothetical particles with very small mass and no charge or spin), neutrinos (particles of very small rest mass, no electrical charge, and that pass through ordinary matter undisturbed).

It is estimated[19] that only 4% of the present universe is known to be atoms of elements classified in the Periodic Table, while 23% is cold dark matter and 73% dark energy. Therefore 96% of the energy density in the universe has never been actually observed, and is only about 1 proton per 4 cubic meters of "vacuum". Although the term "vacuum" or "empty space" is still used, current science (as before noted) supposes

[19] Seife 2005: "What is the universe made of?", p. 78.

that "empty" space at least contains "fields" generated by the presence of material particles that have the four fundamental forces as properties. While after Einstein this "aether" no longer is thought to be the mechanistic aether hypothesized in nineteenth century science, its real existence cannot be questioned without denying the existence of these "fields", an existence empirically inferred from the "Casimir effect", i.e., the mutual attraction of two mirrors in a vacuum.[20] Hence, it is well named simply "plana" (from the Latin adjective for "flat"), as has been suggested by the noted gravity physicist Anthony Rizzi,[21] although others call it "plasma".

Atomic matter is divided into some 117 known elements (a few artificial), 82 of which are naturally abundant. They are classified in the Periodic Table, according to their properties that naturally occur,[22] many with isotopes due to differing number of neutrons in their nuclei. Far the most common elements, as already noted, are hydrogen and helium, followed by oxygen, carbon, neon, iron, nitrogen, silicon, magnesium, and sulfur. Some elements or their isotopes are radioactive and thus unstable, gradually turning into simpler elements.

How Dynamic is Our Universe?

Everything in our universe is in process, although its processes are so interrelated that, for the "present" in which we observe these processes, it is relatively stable, governed by universal laws. All the heavier elements up to iron[23] are forged in the cores of stars with the energy produced by the atomic fusion of hydrogen under enormous pressure, but these elements once produced are eventually scattered in space as the stars which

---

[20] This is a notion not without its problems: see Reucroft and Swain 1998.

[21] Rizzi 2008: *Physics for Realists*, p. 217 note 1.

[22] For various useful forms of this table, see Winters 1993–2010: "The Periodic Table on the Word-Wide Web" <http://www.webelements.com/>.

[23] The suspicion is that explosions from supernova are responsible for elements heavier than iron, such as uranium, gold, etc.

produced them in their death throes explode. Many stars, like our sun, we may expect, will have planets, some with moons, as well as asteroids, meteorites, and comets. Some are variable and others are exploding novae, that is, white dwarf stars in the process of a nuclear explosion due to accumulation of hydrogen on their surfaces.

Our own solar system, besides its relatively large planets, has a broad disk of trans-Neptunian objects in what is called the Kuiper Belt, and supposedly a still more distant Oort cloud of comets.[24] Our system is located half-way from the center of a spiral galaxy, the Milky Way, which consists of about three or even four hundred billion stars, is one hundred thousand light years in diameter and three thousand light years in thickness, although most galaxies seem to contain as few as a billion stars.

Galaxies are elliptic, spiral, or irregular in shape, and may also include clouds of gas called nebulae that may be radiant or reflective, or that block the light of the stars. Some galaxies are called active because they emit large amounts of energy besides that of the stars, probably due to matter falling into the supermassive black hole that seems to be at the core of most galaxies. Galaxies are classified as Seyfert galaxies, quasar, blazar,[25] and radio galaxies. Most galaxies are gravitationally linked to others, some in groups of up to fifty, but others containing many thousands in a cluster. Except for occasional isolated "field galaxies", most are gravitationally bound to a number of other galaxies. Elliptical galaxies seem to be the result of many galactic collisions.

Our own Milky Way is a spiral galaxy, orbited by at least twelve smaller companion galaxies. There are superclusters of

---

[24] We say "supposedly", for the Oort cloud is a hypothetical posit that has so far never been observed. The basis for the posit is the belief by a majority of astronomers that an Oort cloud or something very like it must exist to explain the production of comets.

[25] See *Science Daily* 2009: "Surprising Changes In Black Hole-Powered 'Blazar' Galaxy".

tens of thousands of galaxies that are generally spread out in vast sheets and filaments that surround great voids. No larger structures have been found in our present universe, and thus it seems a relatively homogeneous expanse of supergalaxies and great voids of "plana" in which, however, gravitational and electromagnetic fields exist and through which many forms of radiation are constantly passing. The "empty" space is permeated with a primordial cosmic microwave background which is a faint 2.7 degree glowing remnant from the originally hot big bang explosion. Finally, cosmic expansion due to dark energy is steadily and at accelerating rate driving all these galaxies further and further apart.

These vast structures in space are more finely composed, of course, of atoms and molecules. Some kinds of atoms (such as helium) interact very feebly, but others tend to form more complex molecular wholes. For example, hydrogen and oxygen gases when mixed and ignited become H_2O, the water so necessary for life. Yet, according to our present state of knowledge, *for the most part*,[26] until living substances are present, the molecules throughout the universe remain as relatively simple types.

Moreover, as yet we are not sure empirically that living substances with the power of self-nourishment, growth, and reproduction exist elsewhere in the universe besides on earth.[27]

---

[26] E.g., hydrocarbons, once thought to be the result only of life-processes, have been found now on Saturn's moon Titan, where they appear to have formed independently of any life-processes. Similarly, amino acids have been detected in nebulae.

[27] Though with the discovery that matter is the same throughout the universe, and not divided between an unchanging "celestial" matter and a changing "terrestrial" matter, we know that the capacity of matter to educe ever more perfect forms under the right causal conditions of interaction provides a *strong to overpowering suggestion* that the cosmic development — *from* a universe without life and without capability to sustain life *to* a universe first with regions capable of sustaining life and then, as in the case of our own earth, actually sustaining life and a progressive evolution thereof from lower to higher forms — will have achieved what we find on earth elsewhere in the cosmos as well: see Deely 1969: "The Philosophical Dimensions of the

Modern biology has discovered that more than half the "bio-mass" or living tissue on earth is in the form of minute one-celled creatures that, although they have the vegetative pow-ers of nutrition, growth, and reproduction, lack nuclei and are classified properly as plants. Of true plants, some 270,000 now existing species of green plants and 72,000 species of fungi have been so far identified — and this is probably only about half the real number, thus *vastly* exceeding the hundred-odd named in the Bible, as listed above. Also some 1,324,000 spe-cies of animals (organisms with powers of sensation and mo-tion) have been identified, and this is probably only a tenth of those that actually exist. Yet most of these are insects and other very small creatures. We will say more about these later.

These living things grow on our slightly oblate earth with its oceans and land masses that have a great diversity of moun-tains, hills, plains, forests, deserts, rivers, smaller streams, and lakes, subject to upheavals from earthquakes and volcanoes and the long-range shifting of continents. This scenery pro-vides environments for various kinds of life, from the ocean depths to mountain tops, from the ice caps at the poles to the

---

Origin of Species". Thus Poinsot, writing as a theologian (1643: disp. 42 "De Angelis", art. 1, 474 ¶29), remarks that "whoever wishes to understand the potentiality of matter must also take account of the rational [i.e., intellectual] soul to which matter is in potency, even if it be the case that a rational soul is introduced extrinsically into the material order of creation"; whence (ibid.: 600 ¶71) the appearance of human life in the course of biological evolution on earth belongs to the order of nature and not to any supernatural interven-tion in that order ("etiam anima creatur a solo Deo et infunditur corpori, nec tamen supernaturalis est ejus creatio"), because once the appropriate mate-rial dispositions have been effected in nature "a miracle would be required for God *not* to infuse the rational soul". In Raymond Nogar's summary of "Evolution" (1967: 683): "When the hominid body was so disposed by the natural processes governing the rest of primate development, God created and infused the spirit of man, elevating what was formerly a hominid to the stature of a new, distinct, and unique species" (ibid., 683). We will return to this in later chapters; for now see "2.e. The Intellectual Soul" in Deely 2004: "The Semiosis of Angels", pp. 220–222; and the "Appendix on Immortality" to Reading 3 in Deely 2009: *Realism for the 21st Century*, pp. 87–90.

intense heat of the tropics. To find shelter and food in such variety requires human ingenuity, but the variety surely serves to delight the senses from high noon to midnight stars with immense vistas, awakening our curiosity to explore an ever-surprising world.[28] Geography, geology, oceanography, etc., are branches of natural science that are more descriptive than explanatory, but provide data for natural science in the full sense of the word.

Of the human species there are now some 6.7 billion individuals, and this global population will probably crest at about 10 billion before this century ends. While identical twins or more can arise from a single fertilized cell, this is abnormal and not helpful for the children's development. Moreover, the cytoplasm of the fertilized ovum or zygote contains genomic factors not in the main genome within the nucleus, and hence different portions of this cytoplasm are received by twins in the first cell division of the zygote. Thus the term "identical" is imprecise, and every human person is truly a unique individual. When we also consider that the cortex of the brain of each individual has about 10 billion neurons, most of which have several projecting dendrites each with between 1,000 to 10,000 synapses that communicate with 1,000–10,000 other neurons, muscle cells, glands, etc., we begin to realize that each person is a universe in her or his self. Yet, since there are other planets than earth in our solar system, and since there are at least sextillion stars in the universe, more and more of which we are discovering to have planets, the question arises whether life has arisen there too? and if so, to what organic complexity has it evolved?

Does perhaps the universe also include pure intelligences that, unlike human intelligence, do not require material bodies? Until recent times, in all the major cultures of our planet it was believed by most, including the founders of modern

---

[28] See "To Wonder: Opening the Way of Things", Preamble in Deely 2010b: xxi–xxxvii.

science, that there are such "gods", "spirits", or "angels". These "spiritual creatures", of course, play a very important role in the Bible, although some modern theologians have tried to reduce them to metaphors. The famous *Dutch Catechism*, popular after Vatican II, simply omitted them, until a Vatican commission of Cardinals insisted on their reinsertion. The *Catechism of the Catholic Church* in its 2nd edition gives them considerable attention.[29] It is odd that today scientists who are quite willing, as shown in Chapter 1, to harbor the hypothesis of Many Worlds, have yet closed their minds to the possible existence of non-material substantial realities. Yet, as we argued also in Chapter 1, a critical natural science is logically impossible unless it is founded on a firm empirical basis that includes the certainty of the existence of a non-material First Cause, a Supreme Intelligence who has created the universe in the sense of here and now maintaining its existence outside of nothing.[30] It is this immaterial First Cause that has also created the human intelligence so magnificently exemplified by our great scientists. Yet the human intelligence is the weakest sort of intelligence imaginable, and thus it becomes incredible — or at least strains credulity — that its Creator has not also created greater intelligences than ours, free of the dependency of mind upon body.

Certainly a reason for the reluctance on the part of modern scientists to open-mindedly consider this hypothesis and ask how it relates to their scientific exploration of the universe

---

[29] See #s 57, 148, 311, 326, 328, 329, 331–36, 359–52, 391–93, 414, 525, 538, 559, 760, 954, 1023–29, 1034, 1038, 1053, 1161, 1352, 2676.

[30] It is important to have clearly in mind the point St. Thomas makes concerning creation, namely, that it is not a matter of a first beginning of the universe so much as it is a question of the dependency of finite being as involving potentiality as well as actuality upon a here-and-now source of existence as *purely actual*. See "'Creation' understood as the maintaining here and now of finite existents: the multiaspectual presence of God to the world", in Deely 2010b: 202–205. And see esp. Aquinas' work of 1271: *De aeternitate mundi*.

is that they fear it would detract from their own work: after all, where will such a consideration find its experimental base, outside of "Coast to Coast AM" <http://www.coastto coastam.com/> and anecdotal reports over the centuries of human folklores?

Even so, philosophically, this difficulty in determining an experimental approach to the question no more closes it down than is it necessary to exclude the possibility of extraterrestrial intelligences, "rational animals" that have evolved on some planet or planets other than earth.[31] Indeed, the fact that there is such a Creator who made us in our universe is, as the history of science shows, a wonderful invitation to use the intelligence He gave us to explore in all its details and possibilities His marvelous work of art.

Theologians such as St. Thomas Aquinas have proposed elaborate theories concerning the angels.[32] Aquinas argued that since angels are not individuated by their bodies each individual angel is a species in its own right. Aquinas also considered that their collective number is greater than that of all the species of material living things. If positive scientific arguments can be given for this, as will be examined in Chapter 3, this would mean that the universe is not only quantitatively vast in a material way, but is also metaphysically even greater,

---

[31] Part of the problem is the Enlightenment view of science as exclusively ideoscopic, rather than as presupposing the cenoscopic development apart from which ideoscopy would have no rational foundation, as mentioned in our "Preface" to the present work, pp. vii–viii, and discussed in Chapter 1 above, p. 9ff.

[32] See *Summa Theologiae*, I, qq. 50–64; *Summa Contra Gentiles*, II, qq. 46–55; *Quaestio disputata de spiritualibus creatures* (Disputed Question on Spiritual Creatures), *De substantiis separatis ad fratrem Reginaldum socium suum* (On Separated Substances: for his Companion, Brother Reginald). For translations of these works see Thomas O'Meara, O.P., Bibliography of Thomas Aquinas, at <www.domcentral.org/library/thombibl.htm>. The most important work on angels in the Thomistic line after Aquinas has unfortunately never been translated, John Poinsot's 1643 *Tractatus de Angelis*. Cf. Deely 2004 for work based on Poinsot.

and that the realms of material and immaterial substances are complementary to each other.[33]

We can, in any event, conclude at this point that science has greatly enriched our idea of God and his *power* by showing that, in spite of the fact that its enormous quantitative size has so widely scattered its more complex bodies, the universe contains an immense *variety* of beings. If scientists could see their way to an open mind regarding the possibility of the existence of non-material entities (as they now admit the possibility of embodied extraterrestrial intelligences and multiverses) which by observable material effects could be proved to exist, this enrichment would be wonderfully enhanced. Although it seems that our universe rose from a Big Bang, and perhaps will end in a Big Freeze where nothing much happens, for the present it is dynamically stable enough that scientists have come to study its laws. Those who expand their minds to its message will see in its vast, various, and dynamic mirror the artistry of its Creator, God the Father.

---

[33] See Adler 1982: *The Angels and Us* (though note that this book, while often referring to Aquinas, does not and does not pretend to present the angelology of Aquinas, being far removed in particular when it comes to questions of angelic knowledge and communication). The lengthiest (and most authoritative) Thomistic treatment of angels is that of Poinsot, his *Tractatus de Angelis* of 1643. So far as concerns the doctrine of cognition and communication among angels as derived from Aquinas and Poinsot, see Deely 2004: "The Semiosis of Angels". On the relation in particular of angels to action in space, i.e., in the order of material being, see "How Many Angels Can Dance on the Head of a Pin?" in Deely 2004: 223–227. For a general Neothomistic perspective on the existence of angels see Ashley 2006a: 47–60, "The existence of Created Pure Spirits", Chapter 4 of *The Ashley Reader: Redeeming Reason*.

Chapter 3

CREATION'S ORDER, UNIQUENESS, AND TRUTH REVEAL GOD THE SON

How is the Universe Designed?

For Christian faith Creation is patterned after the Divine Son who is the Word or Pattern of God, as *The Gospel according to John* 1:18 says: "No one has ever seen God. The only Son, God, who is at the Father's side, has revealed him". Thus, conversely, the *Order* or *Design*, *Uniqueness* and *Truth* of the universe, shown us in such detail by modern science, enriches our understanding of God the Son as He reveals God the Father. We must be careful, however, not to make the Pythagorean mistake in seeing this order simply as some kind of mathematical model, a construct, rather than in the full scale of its physically changing, specifically varied, and dynamic character, as described in the previous chapter.

Yet Galileo in initiating the modern Scientific Revolution said:[1]

Philosophy is written in this grand book, the universe, which stands continually open to our gaze. But the book cannot be understood unless one first learns to

---

[1] This quote, one of the initiating statements of the early modern scientific revolution, is from Galileo 1623: *Il Saggiatore*, cited from *The Assayer*, trans. of selections by Drake 1957: 237–238; online at <http://www.princeton.edu/~hos/h291/assayer.htm>. The saying, "Mathematics is the language with which God has written the universe", is often attributed to Galileo, but the attribution is not known to be authentic.

comprehend the language and read the letters in which it is composed. It is written in the language of mathematics, and its characters are triangles, circles, and other geometric figures without which it is humanly impossible to understand a single word of it; without these, one wanders about in a dark labyrinth.

As William A. Wallace OP has shown,[2] one of the basic traditions of human culture is Platonism and Neo-Platonism, often used by the Church in the past to formulate its teachings in terms familiar to educated non-Christians. In Plato's thought, strongly influenced by the mathematics of Pythagoras, reality is understood as a succession of forms emanating from a First Source like light from the sun, as described in Book VII of Plato's great dialogue *The Republic*. At each level of emanation from its Divine Source the light becomes dimmer, but remains to a degree a copy or imitation of all the higher forms as in various degrees they imitate that Source, the One. In the works of the Christian Neo-Platonist, Pseudo-Dionysius (fl. 450AD), this descending-ascending order was given the name *hierarchy* (meaning "sacred order").[3] Thus the universe is seen as a hierarchy of beings whose various forms differ only as grades of one single Source. As shown earlier, this supposition remains the fundamental principle of a materialist like Steven Weinberg in his notion of a "Final Theory" of the universe because he only differs from Plato in not proceeding still further in abstraction to the Ideas that mathematical concepts only reflect. For Plato the *Idea* of 2 is superior to the *number* 2 because there can be many number twos and hence the Idea of 2 is a step from materiality into the level of spiritual beings.

---

   [2] See Wallace 1963, *Einstein, Galileo, and Aquinas: Three Views of Scientific Method.*

   [3] On this, see "Hierarchy in Ecclesiology" in Ashley 2006a: Chap. 12, 171–186, wherein are considered also the views of Alfred North Whitehead. Origins of the term "hierarchy" are discussed in O'Meara 1997. On hierarchy in the Pseudo-Dionysius, see Deely 2001: 132–135.

As we saw, modern science has filled in a very long hierarchy of forms of matter beginning with the various elementary particles. Of these the proton, the oldest and most permanent of these particles, composed of three quarks, is at the center of every atom and hence seems the most perfect and most permanent. The Periodic Table of the elements seems to describe still another level of reality working up from hydrogen to at least uranium. As the atoms form molecules they also can be ordered from simple to complex, with DNA on which life is built at the top of the ladder. Thus living things grade upwards in unified complexity, from one-celled bacteria to plants to animals and then to the intelligent animal, the human person, and perhaps to extra-terrestrial intelligences.

If there is, as scientists suppose, such a hierarchy of material things, why are skeptics not open to the possibility that this is only an imitation of a still more perfect spiritual hierarchy? We have already argued that it is certainly possible that this hierarchy continues, as St. Thomas Aquinas, following Pseudo-Dionysius, describes it, ascending in a hierarchy of pure spiritual intelligences, each one specified by having fewer but more powerful (more "comprehensive") ideas formed under the creative action of God, until the Source of all is reached, God, Pure Thought-Thinking-Itself, *noêsis noêseôs.*[4]

Instead, in current science we normally find *reductionism* by dividing bodies to their smallest parts, i.e., a reverse-hierarchy descending to the material cause is generally preferred. Yet the ascending order enters into science under the term "complexity", and is logically reconciled with reductionism by the claim that material substances differ only (or at least primarily) in the number and variety of their parts, like a machine. Thus the human being is said to be highly complex because of the great number of its genes and the differentiated cells and tissues that

---

[4] Aristotle i.348–330bc: *Metaphysics,* XII, c. 9, 1075b 34.

the genes step-by-step produce (the "phenotype") in interaction with their surroundings.

This reverses the view shared by Plato and Aristotle that what makes one body rank higher than another in the hierarchy of substances is not the number of its parts but its character as a unified whole as it is first observed to exist in full operation. A cathedral differs from a warehouse not primarily in the number or variety of stones and bricks from which it has been built, but in that it is a cathedral, not a warehouse.

Moreover, Aristotle, going beyond Plato, insisted that this superiority should be inferred not from the mere shape in which parts are arranged, but in what through a formal unity the body is enabled to do. "A thing is as it acts". Animals are superior to plants because, like plants, they can take nutrition, grow, and reproduce, but they can also sense food and enemies at a distance and move toward them or escape them. Molecules are superior to atoms not simply because they are atoms bonded together, but because they have energies not possessed by the atoms separately; and atoms are superior to elementary particles because they have properties not possessed by quarks and leptons.

Because Aristotle in his surviving works is so critical of Plato's notion of hierarchy, these works do not emphasize this order of formal causes or "participation" (as it is often called, because the lower shares in the perfections of the higher), but stress rather efficient causality in order to do justice to the dynamic character of reality. Aquinas has been accused of Platonism because he improved on Aristotle's analyses by giving full weight to participation. But basically Aquinas agreed with Aristotle against Plato in *first* giving full weight to efficient causality in its correlativity to formal causality: the efficient cause, as Aristotle saw and Plato agreed, produces the formal cause in and out of the potentiality of matter, although an efficient cause can do this only because it also possesses the same or some superior formal cause. As we have seen, "A thing can give only what it has" is one way to state the Principle of Non-Contradiction.

Aquinas employs the Platonic notion of the "emanation" (flowing out) of forms to signify the way that the properties of a material substance "emanate" from its essence without further efficiency once it has been generated by its efficient cause. Thus a massive body once it exists attracts other massive bodies without requiring any other cause to make it act, and this is true of the other three fundamental forces and of Dark Energy. That is why they are said to be the "fundamental forces" in the universe. Yet the continuance in existence of any material substance (without which continuance, of course, it cannot act) at a metaphysical level still requires the efficiency of the First Cause. Thus Plato's emanation does not adequately explain the coming into existence of a lower substance from a higher. Platonic *emanation* is precisely what Aquinas replaces with his notion of *creation* as the existential dependency of finite being as irreducibly composed of actuality and potentiality upon the infinite being of God as *purely actual* in the most basic and radical sense of actuality, *Ipsum Esse Subsistens*, Existence Itself Subsisting with no intrinsic potentiality at all, with nothing further to be realized on the side of intrinsic constitution.

Of course, one is immediately compelled to ask, "If a thing can give in the order of efficient causality only what it has, then how is an evolutionary universe possible in the first place, since evolution basically consists in the emergence not of lower from higher forms but exactly the reverse, that is, the emergence of higher from lower forms. Does this not violate the Principle of Non-Contradiction?"[5]

There can never be any question of an effect as such exceeding the determination or "perfection" of its adequate reasons for being — a contradiction indeed. But the reason for being is never an efficient cause alone. "There is more

---

[5] "The passage from the lower to the higher grades of being" is "the root problem of evolutionary process" (Deely 1969: 318).

in the cause and the effect than in the cause alone," as de Finance remarks:[6]

> Evolution is a scandal only when the Aristotelian conception of the causal process is reduced logically to the transmission of an identical form, with the individual existence of the effect brushed aside, left entirely outside the account. But in such an accounting, contrary to Aristotle's analytical intention, it is the truth itself of efficiency which one compromises.

To challenge the idea of an evolutionary universe, such as modern science has made to be unmistakably the physical context of life on our planet, and of our planet itself within the stellar context of the Milky Way and surrounding galaxies, is to make no allowance for the pre-existence of the "patient" and the repercussions of its own pre-existing organizational dispositions which may either reinforce or cancel out or modify in some startling way the dispositions which *would have been* established by the efficiency of the agent *if* its interaction partner had been a purely plastic material. Since "the corruption of one form is the generation of another", and since all forms are corrupted only *per accidens*, it is to the final dispositions of the being corrupted that we must look if we wish to know the actual species of the subsequent form.

This is clear from Aristotle's definition of the soul (the substantial form of a living being) through its proper subject: "If, then, we have to give a general formula applicable to all kinds of soul, we must describe it as the first grade of actuality of a natural organized body",[7] where the "natural organization" in question is simply the microstructural dispositions which will

---

⁶ Joseph de Finance 1955, *Existence et liberté*, p. 262–263. For example, this seems to have been the basis for Hegel's express denial of the possibility of an historical evolution such as Darwin argued for: G. W. F. Hegel, 1830: *Enzyklopädie der philosophischen Wissenschaften*, p. 202, par. 249. See the comments on this passage in W. T. Stace 1955: par. 434, pp. 313-315.

⁷ Aristotle c.330ʙᴄ, *On the Soul*, Book II, ch. 1, 412b4–6.

necessitate the eduction from matter of a form with the faculty of reproducing — matter organized in such a way as to enclose within the substance generated the capacity for life. For "unity has many senses (as many as 'is' has), but the most proper and fundamental sense of both is the relation of an actuality to that of which it is the actuality",[8] as the pupil plus the power of sight constitutes the eye.

The whole question turns on the problem of organization. The total range of diversity in the universe of physical beings is rooted in the peculiar disposition and composition of parts in each unity, that is, in the individuating disposition; but because there are in the universe as we know it four irreducible levels of material existence (i.e., inorganic matter, plant life, brute animal life, rational animal life), this individuating disposition must also always include a specifying disposition.

Living bodies, as all natural bodies, are fashioned out of pre-existing matter, i.e., out of the potentiality in each thing to be converted, remotely or proximately, into something radically different. Thus, considered in itself, life pertains to the potency of matter. *Per se*, the organization specific of life (realizable according to countless concrete modes) belongs to inorganic matter only after the manner of an inadequate or remote potentiality; *per accidens*, however, it may under given conditions pertain to it adequately, i.e., causally.

This is the basis for the prior possibility in principle of so-called "equivocal generation": the origin of a "higher" form replacing a "lower" one — the origin of living matter out of non-living matter by reason of a fortuitous dispositioning of the latter in a chance series of causes.[9] That this is possible follows

---

[8] *Ibid.*, 412b8–9.

[9] In the evolution of the universe from its original preliving condition to a condition capable of and then actually sustaining life, we see exhibited unmistakably the role of chance at play in a universe freed from the myth of the unchanging heavens controlling generations and corruptions upon earth: the chance event lacks an intrinsic determinism, but it is nonetheless *contextually* determined by the overall pattern of substantial interactions which make

from the very nature of the soul understood as a substantial form, i.e., the first act of a body disposed through organization to sustain in being the operations of life. It does not matter by what agencies this organization is effected: the sole condition essential and primary for educing a soul (= for educing the substantial form constituting a living being) is the production of an organization suited to life; the actual processes through which this organization is constituted are accidental and purely secondary considerations.

A *univocal cause* is always proportioned to its effect, either in the sense of belonging to the same irreducible level of reality, or in the sense of belonging to a higher order, such that it contains the real species of its effect within itself eminently. An *equivocal cause*, on the other hand, need not be proportioned to its effect except *per accidens*, in the general way that any material substance is able to act on another by very reason of belonging to a common genus within reality. In this way, as the modern investigations of biochemistry sufficiently indicate, the structures of the living world are potentially latent throughout the whole of secondary matter (that is, the entire realm of what can be seen and touched, whether directly or through the use of instrumentation);[10] for which reason again a concatenation

---

up the physical environment. See "Chance Events", in Deely 2001: 66–67; further discussion in De Koninck 1935: *Le Problème de l'Indéterminisme* and 1937: "Réflexions sur le Problème de l'Indéterminisme", in Deely 1969: "The Philosophical Dimensions of the Origin of Species", and in Dennis Bonnette 2003: *Origin of the Human Species*, 52–55, 68–71. But note further that the same principles which explain the possibility of life originating from interactions of lifeless matter would apply to a laboratory-controlled causal series. And while the human case has a distinctive uniqueness not found even at the level of animal origins in general, the point that Poinsot makes in his *Cursus Theologicus* (cited in Chapter 2 above, page 56 note 27) remains, namely, that human origins pertain nonetheless to the order of nature.

[10] E.g., consult N. H. Horowitz 1956: "The Origin of Life"; S. Huang 1959: "Occurrence of Life In the Universe"; Albert Ducrocq 1957: *The Origins of Life*. Two works in this area are fundamental classics: A. I. Oparin 1957: *The Origin of Life on Earth*, and L. J. Henderson 1958: *The Fitness of the Environment* (Boston: Beacon, 1958). See also Gredt 1961: *Elementa Philoso-*

of special circumstances could efficaciously, though in a *per accidens* way, dispose the really specific organization of a living being which otherwise pertained to any one of the circumstanced entities only potentially and indeed inefficaciously. "It is evident", De Koninck notes,[11] "that every natural generation involves a measure of uncertainty", which could not be the case were the form from the efficient cause the sole determining factor of the effect:

> And therefore when a superior nature is produced from the potency of an inferior nature by equivocal generation, this production remains natural, obviously not in relation to the inferior agent considered as inferior ... but to the extent that it responds to the desire of the inferior nature *in so far as* it is ordered to the good of universal nature, and to the intrinsic last end of the world. And if the consideration of the hierarchy of cosmic species is an '*opus naturae*', the irregularity of this gradation, such as is apprehended by experimental science, is sufficiently explained.

Thus, in the case of the evolution of life, even when there is a passage to a higher level, there would be no violation of the principle of causality, and no need for a "special" divine concursus (still less intervention), any more than there are instances of either of these in our everyday experience.[12]

The soul, after all, like every substantial form, organic or inorganic, is but the *first* actuality of a disposed physicochemical structure. From an *experimental* point of view, it is only a superficial difference in kind:[13]

phiae Aristotelico-Thomisticae, Vol. 1, footnote 1 p. 349, and Section 408, pp. 342–343.

[11] De Koninck 1937: "Réflexions sur le problème de l'indeterminisme", p. 238.

[12] As again De Finance remarks in the text cited in note 6 of this chapter, p. 66 above.

[13] Sertillanges 1945: *L'idée de création*, p. 147.

Since in the two cases the empirical conditions are different, the phenomena themselves differ: that is all that the scientist finds.

From a point of view *explanatory* of the real, it is a radical difference in kind, an irreducible level or zone which will fill itself up with novelties until the opening of a still further zone is required by the very exuberance of the vegetative forms. Thus will the universe pass, by reason of the "unintended consequences" (i.e., the chance effects) within the finalities of its interactions at the level of efficient causality from a universe lifeless and incapable of supporting life to a universe first capable of and then actually supporting life, eventually to include human life on our planet and perhaps elsewhere[14] (as will eventually become known if we "live long enough" as a scientifically inquiring species!)

Like De Koninck's imaginary intelligence contemplating the earliest stages of the process, we recognize the inevitability of life once constituted to rise — barring catastrophe — by the steps of historical novelty and by multifarious and weaving paths leading through a maze of natural forms through a taxonomic hierarchy to the rational animal; and, within the 1,600,000 plus types of animals (including here the 800,000 plus types of insects which in an ontological no less than in a strictly biological optic are modes of animality: *sentire in sentibus est esse* — "to sense in sensory beings is to exist") and 200,000 plus types of plants called by the taxonomist "species", we recognize another order of species and another hierarchy, the perfect hierarchy of essential forms.

We see, therefore, how the problem of the "higher" from the "lower" poses itself within the order of ontological grades.

---

[14] "The natural species, which are quasi-genera in relation to the subspecies, are certain *a-priori*, because they constitute irreducible degrees of being: there is nothing intermediary between 'being', 'living', 'cognizing', and 'understanding'" (De Koninck 1937: "Réflexions sur le problème de l'indeterminisme", p. 234).

It is a mistake and a complete misunderstanding to state the issue in terms of dogs generating humans or butterflies generating mice. The authentic philosophical question is whether there is some form of physico-chemical organization which could under some circumstances be so disposed by the cosmic agents environing it as to require the eduction of a living form; and beyond that a question of whether there is any form of vegetative life which could under some circumstances give rise to some form capable in however imperfect and rudimentary a way of sensitive life. And from the standpoint of the definition of the soul through its proper subject and the involution of the causes, it is impossible to say that an affirmative answer to this question involves a contradiction. By reason of the fact that the real species can only be realized in individuals historically and contingently constituted, it is impossible to assign to these species absolute limits. "Species are ordered hierarchically," Adler notes,[15] "but within each sphere of the hierarchy — which, because of the intrinsic indetermination and contingency of these forms is a zone of probability — there is a continuum of sub-species, varieties, or races which are only 'statistical entities'."

Given the presence of matter and the lability and indeterminacy thereby imparted inseparably to forms, the interactions of finite beings inevitably gives rise to a *progression* from "lower" to "higher" in the course of the efficient causalities exercised over the natural course of generations and corruptions under the influence of the First Cause as sustaining in existence (the "creation out of nothing") of the universe of finite beings interacting.

Exactly as St. Thomas expressly remarked, the discovery that the universe is as changing throughout as it is on earth (that there are no "celestial spheres") would lead inevitably "to some other explanatory scheme not yet conceived of by

---

[15] Adler 1940: *The Problem of Species*, p. 82.

men",[16] specifically, as we now realize, from a static to an evolutionary view of the universe as a whole.

Why is God the Son the Word?

"In the beginning was the Word, and the Word was with God,
and the Word was God.
He was in the beginning with God. All things came to be
through him,
and without him nothing came to be.
What came to be through him was life,
and this life was the light of the human race;
the light shines in the darkness, and the darkness has not overcome it"

— *John 1:1–5*

This revealed truth utterly transcends human reason and science, but has been made obligatory for us to believe on faith because of the signs that manifest it. A principal such sign evident today, for those who will examine it open-mindedly, is, as Vatican I defined[17] and as confirmed by Vatican II in the *Dogmatic Constitution on the Church*, the Church herself, because she is:[18]

> By her astonishing *propagation*, outstanding *holiness*, inexhaustible fertility in every kind of *goodness*, *catholic unity* and unconquerable *stability* a kind of great and perpetual *motive of credibility* and an incontrovertible evidence of her own divine mission.

Here, by "motive of credibility" is meant (within the context of orthodoxy)[19] "the reason that one has a rational obligation

---

[16] Aquinas c.1272/73: *In libros de caelo et mundo*, Book II, lectio 17, n. 451. See, in Deely 2001: 263–266, "Cosmology in Aquinas".

[17] Vatican I 1870: Session 3, *Dogmatic Constitution on the Catholic Faith*, Chapter II, 12.

[18] Vatican II 1965: *Lumen Gentium*, Dogmatic Constitution on the Church; see *Catechism of the Catholic Church* n. 156.

[19] This is an important qualification. Thomas Aquinas, writing around

to believe", so that to refuse to accept and trust what apostolic authority says risks to deliberately deny the truth.

The notion of participation or hierarchy of formal causes is used theologically to provide an explanation of why the Divine Son is called by St. John the *Logos* or Word. This term was much used in the Stoic philosophy founded by Zeno of Citium (fl. 300BC), who was much influenced by Aristotle but was himself a materialist, to designate the "fire" or energy which causes law-determined motion in the universe. Whether this Stoic view, widely spread in the Roman Empire in St. John's time, influenced the fourth evangelist's use of the term is uncertain, but he uses it in the sense of God's creative "word" or command. ("Then God said [Gn 1:3], 'Let there be light', and there was light") that gave order to creation. This sense of *logos* is thus quite different than what the Stoics and today's pantheistic and materialistic "cosmologists" signify by the term. Yet the analogy of light (fire) and its degrees is included in the biblical notion of God's Word as a revelatory shining forth.

Thus the modern emphasis on light as the highest form of energy that reveals to us the structure of the universe enriches the theological notion of the Son as the pattern or form after which all creation, and especially that of intelligent creatures, is brought into existence from nothing. "So God created human beings in his own image, in the image of God he created them; male and female he created them" (Gn 1:27). Aquinas,

---

1264 on "how to argue with unbelievers", warned not only that one "should not try to prove the Faith by necessary reasons", but warned also that to pretend in argument or to convey by our attitude that such proof is available, is to make a mockery of the very faith one professes. Aquinas c.1264: "On Reasons for Religious Belief", Chap. 2; see also 1266/73: *Summa theologiae* 1.32.1c, and 1265/66: *De Potentia Dei* q. 9. art. 5. Thus his brother in religion of the late 20th century, Vincent Guagliardo, OP (2011), accurately summarizes Aquinas' view this way: "The theologian, as one who stands within faith (and not above it), is not in the position of 'proving' faith from some privileged vantage point, but only of educing arguments which show the plausibility, or reasonableness, of faith, i.e., that what is proposed by faith is not irrational or contradictory to reason."

therefore, distinguishes between a superior participation in God which is an "image", and a lesser "trace" (*vestigium*) or likeness found in sub-human creatures.[20] The more science reveals to us the hierarchy of forms in the universe, the brighter shine these images and traces that are within our experience.

The advance of science is often intertwined with that of the fine arts. The Catholic Church has always encouraged the fine arts that teach us to observe the still greater cathedral of the creation that reveals its Maker and of the human body (and especially the human face) that reveals the inner beauty of the human spirit as well. The heresy of iconoclasm and its revival in Protestantism, Calvinism particularly, was a great mistake, although the Protestant emphasis on preaching and the beauty of the Biblical Word somewhat compensated for its minimization of the religious value of the sacraments and of art.

As Protestantism gave way to Secular Humanism a new kind of iconoclasm appeared, after Impressionism, in the form of "abstract" or "non-objective" art seeking to minimize or even eliminate the "imitative", "objective", "representational" aspect of painting and sculpture that Aristotle in his *Poetics* praised as *mimesis* or imitation and argued to be the very essence of fine art. In the interest of enhancing the subjective expressiveness of the artist, non-objective art reduces representation and emphasizes the artist's control, as in cubism, or the artist's spontaneity, as in the "drip paintings" of Jackson Pollack. This fits the religion of Secular Humanism in which the artist is a "creator", but there is no Creator.[21]

In making this criticism we have no intention of denying the greatness of much modern art. Protestant subjectivism, with its emphasis on "faith" rather than on "what one has faith in", joined with Cartesianism to produce the subjectivism of idealist philosophy and this has ended in so-called "analytic

---

[20] Aquinas 1266: *Summa theologiae* I, q. 45, art. 7c; q. 93 art. 1–9.
[21] See Ashley 1965: "Significance of Non-Objective Art".

philosophy" that is little more than grammar. Yet idealist philosophy has called our attention to one set of strands in what has been described as "the semiotic web"[22] and thus can be of great value for future thought. Modern art has also produced masterpieces that academic art with its photographic representationalism could not have created. In particular modern art has cultivated so-called "formalism", or insistence on the relationships of figures and colors without any resemblance to nature, and this has produced genuine masterpieces.

Yet pre-modern representational art was also very aware of composition and color balance, and surpassed mere photography. In fact modern photography is also artistic in its selective formality. Aristotelian *mimesis* is not photographic representation, since he wrote in *Poetics*, chap. 9: "Poetry, therefore, is a more philosophical and higher thing than history; for poetry tends to express the universal, history the particular" (note that for Aristotle *historia* means a collection of factual data). Thus mimesis imitates not merely by copying but by selecting those actions of the characters that illustrate some essential truth about human life. Visual mimesis, therefore, selects elements of what the artist sees or imagines that reveal the essential natures of things by a kind of abstraction or emphasis. A Chinese artist once said that, to paint a bird, he observed it for many days, and then, without looking at it, drew its picture, thus catching what was most characteristic in its appearance and behavior.

It might be objected that many natural things, especially animals, are grotesque or positively ugly in the eyes of many, such as an alligator, porcupine, or rhinoceros, and especially the ancient dinosaurs. Something will later be said about these "monsters" in discussing the evil in the world, but here it is sufficient to point out that their ugliness is such only in comparison to especially beautiful objects that they

---

[22] See "The Semiotic Web" in Deely 2001: 605–606, and the Index entry "semiotic web", pp. 989–990. Re analytic philosophy, see in particular Deely 2006.

somewhat resemble. Yet genuine artists find much beauty in every living thing, as well as in inanimate objects. A mountain disrupted by volcanic action displays a new, though awesome, magnificence.

As artistic mimesis seeks the universal and essential in history, so science seeks to do in a still more abstract manner, since it attempts to find the form or structure of reality by leaving behind everything accidental and formulating essential relations, often in mathematical formulas, admired by scientists for their beauty or elegance. In a well known quote Bertrand Russell wrote:[23]

> Mathematics, rightly viewed, possesses not only truth, but supreme beauty — a beauty cold and austere, like that of sculpture, without appeal to any part of our weaker nature, without the gorgeous trappings of painting or music, yet sublimely pure, and capable of a stern perfection such as only the greatest art can show. The true spirit of delight, the exaltation, the sense of being more than Man, which is the touchstone of the highest excellence, is to be found in mathematics as surely as in poetry.

Such elegant structures can also be ordered hierarchically in order of complexity and unity. Yet in doing so we must keep in mind that, as mentioned earlier, even mathematical concepts, because they are quantitative, retain a reference to potentiality, to infinite divisibility, and hence are by no means absolutely "clear and distinct" as Descartes claimed. This is especially true when continuous quantities are reduced to discrete quantities, or vice versa, and then, for the sake of still greater generality, analogies are introduced and thus paradoxes seem to arise (as in the case of "infinite numbers").

While the most evident form of the beautiful is visual or auditory, literature, whether poetry or prose, more perfectly conveys universal truth. That is why God wrote the Bible in

---

[23] B. Russell 1918: "The Study of Mathematics".

human language. It is also, however, why biblical fundamentalism is a misguided understanding of biblical revelation. The various human authors of the Sacred Scriptures differ in their literary skills and in the particular literary forms they use, yet all speak the Word of God. Jesus in his parables and discourses, St. Paul in the various tones and formalities of his epistles, St. John in his Gospel and letters, the *Book of Revelation* and "Solomon" and the Prophets in their preaching, used a variety of styles. Theological exegesis, therefore, must take all of this into account, as the Fathers of the Church in their commentaries sought to do, and as Galileo in his 1615 *Letter to Grand Duchess Christina* quite ably did (though in vain,[24] as far as Cardinal Bellarmine and the Roman Inquisition were concerned) in defending himself against the accusation that heliocentrism contradicts the Sacred Scriptures. The Galileo Affair, however, also illustrates the hermeneutic caution required by the writings of scientists. They can range from mathematical formulas so general that they touch reality only at a few points to enthusiastic science fiction such as Thomas Kuhn exhibited by his concept of a "paradigm shift". We should not, however, fall into the error of thinking there is no permanent truth.[25]

We can, therefore, conclude that science, by pointing out what is essential in the great variety of objects in our universe, dynamically interrelated by fundamental forces, also finds in them a hierarchical order, from the simple to the complex, and unified in a way in which mathematical order is given a concrete but ever-changing realization.

In current science the classification of things, however, tends to remain reductionist — that is, conceived mainly in terms of the material cause. Carolus Linnaeus (1707–1778) founded modern biological taxonomy principally on anatomy,

---

[24] Recall our discussion of the "larger point" of the Galileo case in Chapter 2 above, pp. 41–42, note 10.

[25] Recall Weinberg's 1998 analysis of Kuhn's work under the title "The Revolution that Didn't Happen"!

that is, on the parts of a living whole; and this still survives in current classifications based on genetic DNA. Aristotle, on the other hand, while not denying the importance of the material cause, and using it in *De Partibus Animalium*, held that the superior criterion would be the formal cause, that is, the *behavior* of the living thing, because "a thing is as it acts".

In the case of animals Aristotle's criterion would be best applied by studying the psychology of the animal and determining the rank of its sense powers, especially of the internal senses. Yet at present this has proved very difficult. For example, chimpanzees in some respects seem to have greater learning abilities than gorillas in tool making, but less in developing communicative skills,[26] although this is disputed.[27] Thus biologists, when they try to mark off species by behavioral criteria, generally do so by the inability to reproduce fertile offspring, which is more a classification by vegetative than by animal acts. In Chapter 4 we will see how this current taxonomic inadequacy affects the theory of evolutionary descent.

---

[26] Not to be confused with species-specifically human linguistic communication, as Sebeok has definitively established in numerous studies, both of his single authorship (1963, 1968, 1975a, 1977, 1978, 1984, *inter alia*) and in collaborative works, notably Sebeok ed. 1968, Sebeok and Rosenthal eds. 1981. Sebeok's demonstration of the distinction between linguistic communication as species-specifically human and generically animal communication is synthesized in Deely 2007.

[27] See Francine Patterson and Eugene Linden 1985: *The Education of Koko*, Chap. 3; see further <http://www.primatesworld.com/TalkWithChimps. html>, also <http://whyfiles.org/058language/ape_talk.html>, for further illustration of this line of research into "animal communication". In general, as Sebeok and others (preceding note 26) and others have amply demonstrated, the various so-called "ape language" studies are informative yet argue repeatedly, using inadequate criteria, that it is not yet possible to give a conclusive answer to the question of "talking animals" since, for example, chimp behavior is very close to being human. Overviews of current discussion are found in the *Wikipedia* entries "Animal Communication" <http:// en.wikipedia/org/wiki/Animal-communication> and "Animal Language" <http://en.wikipedia.org/wiki/Animal_language>.

Aristotle, in his *De Partibus Animalium*, pointed out that the classification of animals by genus and species can be ambiguous by reason of what modern evolutionists have come to call "convergence". For example, whales, although they swim like fish and have fins, are in fact mammals with very different structure from a fish and certain behaviors quite unlike a fish. Aristotle also mentions border-line cases, like the sponges that appear to be motionless like plants, yet dilate and contract their openings like an animal. In a hierarchy, the "highest of the lower is like the lowest of the higher".[28] Yet this generic ordering leaves something to each species *not* contained in any of the higher species except the First Cause.

Is Each Individual Creature Unique?

While in a Neo-Platonic hierarchy each member of the species contains *all* the forms of lower species, in Aristotelian-Thomistic hierarchies each species relates to the First Cause both generically and specifically, but the specific participation is the more profound. Thus, while we human beings are superior to minerals, plants, and brute animals, we are so generically even though we are not more yellow than gold, taller than oak trees, or more volatile than a bird. Even in the hierarchy of the angels as St. Thomas explains it, in which linearity of rank is most perfectly realized, each angel is a unique species, although this is caused, not by matter and quantity as in bodies, but in the freedom by which each angel thinks for itself and loves God in a special way. Human persons, as the border-line between materiality and spirituality, are first individuated by their separate bodies, and only then by their conscious and free "personalities" as well.[29]

---

[28] A relevant application of this principle can be seen in Deely 1971: "Animal Intelligence and Concept-Formation".

[29] Aquinas 1266: *Summa theologiae prima pars*, q. 90, art. 4, responding to the assertion that "the human soul is more like the substantial form of angels than that of brute animals", answers to the contrary that "if the human

Aquinas argued that God never makes a mere copy of
something, but that each being, even a grain of sand, has a
uniqueness, a formal element that it shares only with its Creator.[30] At first sight this seems to be contradicted by current
quantum theory, in which each elementary particle is identical with other members of its species, so that all electrons are
exactly alike. This, like claiming that an electron is merely a
point, however, confuses a mathematical description with the
reality it models, reducing the "reality" to a mathematical construction; it cannot be taken strictly, or it would be a contradiction to speak of two or three electrons.

This problem arises also in the "Many Worlds" or "Multiverse Theories" mentioned in Chapter 2. If these worlds in no
way affect each other it is pointless to talk about them scientifically, since they can neither be observed in themselves nor in
their effects in our universe. To call something "possible" an
instance of which cannot be observed is futile, a "meaningless
surplussage" as Peirce said of the Kantian notion that things
in themselves are "unknowable",[31] since a merely hypothetical

---

soul by itself were a natural kind, it would indeed be more like the forms of
angels than like the forms of brute animals; but because it is the form of a
body, the human soul achieves the speciation of a natural kind only as formal principle of an animal body, and therefore within the genus of animal";
wherefore the human soul is not properly conceived as the embodiment of a
spirit but rather as the principle whereby the genus of animal itself becomes
spiritualized through the human species of animal. See Deely 2011: "Toward
a Postmodern Recovery of 'Person'."

[30] On this see "Hierarchy in Ecclesiology", in Ashley 2006a: *The Ashley
Reader*, Chap. 12, pp. 171–183.

[31] "In half a dozen ways", Peirce comments (c.1905: CP 5.525), the Kantian notion of an *unknowable* "Ding an sich has been proved to be nonsensi-
cal; and here is another way", he adds: "It has been shown [CP 3.417ff] that
in the formal analysis of a proposition, after all that words can convey has
been thrown into the predicate, there remains a subject that is indescrib-
able and that can only be pointed at or otherwise indicated, unless a way of
finding what is referred to, be prescribed. The Ding an sich, however, can
neither be indicated nor found. Consequently, no proposition can refer to it,
and nothing true or false can be predicated of it. Therefore, all references to

reality may always contain an inner contradiction that we have not noted, because we know nothing real exhaustively. Yet if these universes act on each other in some way, then they form a super-universe that we *can* in part observe. Then, because we cannot say scientifically — *pace* Hawking and Mlodinow — that this "just happened" to be, we are back in the problem of the First Cause, since we cannot say that anything that is changing intrinsically is *causa sui*, as already shown in Chapter 1. Even if this hypothetical super-universe contains an infinity of possible universes, the same need of a First Cause holds, and all that is achieved is to exalt its own infinity. In fact, at present the only evidence for a Multiverse are the problems of giving an interpretation to the mathematical Standard Quantum Theory, a common crisis in the development of natural science in which new dilemmas constantly arise.

An oft-repeated historical error is the assertion that Aristotle and medieval thinkers wanted a geocentric astronomy in order to enhance human dignity by putting themselves at the center of the universe. In fact, in the old astronomy the earth was at the center because the surrounding heavens were composed of a "quintessence" (a fifth kind of matter, in contrast to the four earthly elements involved in substantial generations and corruptions) rotated eternally by bodiless, spiritual intelligences, and immune to any other kind of change than their ceaseless circular motion and emanation of light along with various astrological influences. By contrast, within the "sphere of the moon" centering on the earth, matter was deemed to consist of four kinds of atoms (dry-cold earth, wet-warm water, wet-cold air, and dry-hot fire) in constant process of altering each other and moving randomly upward or downward. We humans, therefore, by reason of our bodies composed of

---

it must be thrown out as meaningless surplusage." See "The First Possibility", in Deely 2008: 103–105. For a full discussion of Kant's noumenon/ding-an-sich distinction may consult Deely 2001: "The Synthesis of Rationalism and Empiricism", 553–572, esp. 558–559.

these elements, yet having spiritual souls, were the "highest of
the lowest, but the lowest of the highest" creatures. Stretching
far above us in the hierarchy of beings were the vast celestial
spheres and the angels moving them, these angelic beings existing, according to Aquinas, in greater numbers than all terrestrial species. Thus, in this ancient and medieval Aristotelian
astronomy, we were quite humble beings, and being "at the
center of the universe" was more a disgrace than an honor.[32]

Strikingly to the contrary of this ancient view, the view that
began to take hold in the Renaissance vastly elevated human
dignity, and today there is no direct evidence that anything
in our universe is as complexly constructed or even as intelligent as the human species, formerly thought to be located so
lowly (i.e., *near* the center, Hell being the *actual* center!). It is
true, of course, as already mentioned, that there are interesting
probable arguments (but as yet no direct evidence) for either
pure spirits (angels) or even for extraterrestrial intelligences
embodied on other planets so often pictured in science fiction
and sought now over that last half-century by SETI.[33] Their
advocates hypothesize (the "Drake Equation")[34] that, by this

---

[32] Thus Dante, for example, located hell at the earth's center, that being
the very farthest remove from the lofty spheres of heaven and hence the most
ignoble location in the entire physical universe.

[33] See <http://www.space.com/searchforlife/>.

[34] This equation, developed by Frank Drake in 1961 "as a way to focus on
the factors which determine how many intelligent, communicating civilizations there are in our galaxy", reads as follows: $N = N^* f_p n_e f_l f_i f_c f_L$, where
to the left of the "equals" sign "N" is the number of communicating civilizations in the galaxy, while to the right of the equals sign are the determining
variables to be multiplied together. "N^*" represents the number of stars in
the Milky Way Galaxy, "f_p" is the fraction of stars that have planets around
them, "n_e" is the number of planets per star that are capable of sustaining
life, "f_l" is the fraction of planets in n_e where life evolves, "f_i" is the fraction
of f_l where intelligent life evolves, "f_c" is the fraction of f_i that communicate,
and "f_L" is the fraction of the planet's life during which the communicating civilizations live (using our own case as an example, if we suppose that
our civilization endures for 10,000 years, f_L in that case = 1/1,000,000). See
<http://www.activemind.com/Mysterious/Topics/SETI/drake_equation>.

time, on many planets intelligent life must have evolved, and perhaps far beyond the level that our poor human intelligence has achieved. Three questions are raised by this hypothesis: (1) the famous "Fermi Paradox": "Where is everybody?" — that is, why haven't the ETs communicated with us — that has led to all the unproven UFO sightings; (2) the fact that we have so far found no life on other of our own planets (i.e., on other planets in our solar system) makes evident how often a planetary system can exist without life, so that the Drake Equation has been reduced by newer calculations from Drake's 1.00 to 0.08, that is, less than 50% chance; and (3) even if there exist ETs, can their brains really be much better than the human brain, which seems near the limits of organic complexity?[35]

In any case the modern view of the universe, by comparison with the ancient or the medieval view, far from *reducing* human dignity, *raises it immensely* by reason of the Anthropic Cosmological Principle that states that our universe must be as vast and ancient as it is or intelligent life could never have evolved within it.[36] This view which is accepted by many leading scientists — while even those who, like Steven Weinberg, have belittled it,[37] and yet use it to bolster the acceptance of the Many World Theory[38] — has a weak and a strong form. The weak form (that Weinberg favors) simply states that our existence seems very improbable since it depends on the coincidence of a great many factors any one of which, if only a little altered, would have blocked human evolution. The strong form is the Design Argument that concludes from this very "improbability" that an intelligent Creator must exist who

---

[35] See Merkle 1989: "Energy Limits to the Computational Power of the Human Brain", and 1988: "How Many Bytes in Human Memory"; also Ward 1997: "End of the road for brain evolution".

[36] Barrow and Tipler 1986: *The Anthropic Cosmological Principle*.

[37] See Weinberg's 1999 article, "A Designer Universe", cited in Chapter 2 above at note 22, p. 15.

[38] See Weinberg 2005: "Living in the Multiverse".

has designed the universe to produce our planet and the evolution of intelligent life. Indeed, those scientists who, like Carl Sagan,[39] seek to communicate with ETs superior to us in intelligence, promote hopes of space travel and of the invention of artificial intelligence that ascribes to the human mind powers to understand and control the universe that ancient and medieval thought would have considered wildly extravagant.[40] One might suspect that such exaggerated views of human intelligence and scientific-power-without-limit are substitutes for belief in a higher spiritual reality by scientists. Monotheistic theologians, however, need not depend on the Design Argument as the primary rational proof of God's existence,[41] for they have the much more basic First Way based on the simple fact of change, as we saw in Chapter 1. They know with scientific certitude that God is Existence Itself and First Cause of all finite being (including the multiverse if there is such a thing!).

---

[39] Sagan, in the 1980 *Cosmos* series, questioned the practicality of interstellar travel and the possibility of UFOs visiting Earth. Yet he also wrote that the stars "beckon" to humanity, and thought that the Bussard ramjet might attain interstellar travel. In one of his last written works, *The Demon-Haunted World: Science as a Candle in the Dark* (1986, pp. 81–96, 99–104), he still denied that there is hard evidence for UFOs, past or present. See the entry "Carl Sagan" in *Wikipedia* <http://en.wikipedia.org/wiki/Carl_Sagan>.

[40] Of course, the medievals would have considered wildly extravagant the degree to which we can now anticipate the course and development of hurricanes! In general, even though the medievals accepted Aristotle's idea that our "speculative understanding of the way things are becomes practical by extension", they never dreamed of that extension reaching so far as to make humans responsible for the well-being in general of all the earthly life forms! See Petrilli and Ponzio 2002: *Semioetica*; also Deely, Petrilli, and Ponzio 2005: *The Semiotic Animal*; and "The Ethical Entailment of Semiotic Animal, or the Need to Develop a Semioethics", in Deely 2010a: *Semiotic Animal. A postmodern definition of human being transcending Patriarchy and Feminism*, pp. 107–125.

[41] And fortunately so, since an apodictic "design argument" is neither easy to come by nor found among the "five ways" as summarily presented by Aquinas in his *Summa theologiae* of 1266 at I, q. 2, art. 3. See Christopher Martin 1997: *Thomas Aquinas. God and Explanations*, Chap. 13, esp. pp. 182–201; and "The Deficiency of the 'Fifth Way', and the Matter of Alternative Further 'Ways'," in Deely 2010b: 190–191.

The constantly increasing evidence of the Anthropic Principle in its weak form is only a secondary reason for "I believe in one God, Creator of heaven and earth".

We can conclude, therefore, that the theological contemplation of God the Son as the Word made flesh in Jesus Christ is enriched by the scientific evidence of how human persons, associated perhaps with other embodied but extraterrestrial intelligences who parallel us on other planets, are the summit of the material universe, created in the image of God. It is by analogy to such persons as we can observe them that we can understand something of their Creator. Hence scientific anthropology proves essential to the development of Christian theology.

How is the Son of God Truth Itself?

The success of science in retaining the fundamentally empirical way of thinking defended by Aristotle and Aquinas against Plato, Descartes, and Kant helps us to understand why God, who is Truth Itself, chose to reveal that truth to us according to our human modes of thought. Ultimately He did this through the Incarnation of his Son, the True Word Itself, as one of us, Jesus of Nazareth. "And the Word became flesh and made his dwelling among us, and we saw his glory, the glory as of the Father's only Son, full of grace and truth" (Jn 1:14). "For we do not have a high priest who is unable to sympathize with our weaknesses, but one who has similarly been tested in every way, yet without sin" (Heb 4:15). This revealed truth places the species to which human nature belongs above every other species in the universe, not because it is as such superior, but because it has been assumed by the Divine Son precisely because of its lowest place in the spiritual realm of the universe, although highest of material beings. It is the highest material being with a spiritual form, but the lowest of all spiritual forms, and is thus at the border-line between the scope of natural science and that of metaphysics. Since natural science is from sense knowledge to this border line of spiritual knowledge, it cannot help but enrich our theological understanding, centered as it also is on the Incarnate Word, Jesus the Nazarene carpenter, still present with us in the Eucharistic bread and wine.

Today this border-line puzzle, called the Mind-Body Problem, is widely debated by scientists but generally given by them a simplistic materialist answer. Many scientists confidently suppose that, in the near future, they will be able to produce artificial computers that will out-do them in "thinking". The very term "Mind-Body" shows the deep influence of René Descartes' *dualism*, and behind it that of Plato, for whom the human intelligence was immaterial and eternal and, as previously noted, able to think by recalling the innate ideas which it had eternally received by emanation from the One.[42] On this Platonic view, the human soul descends into the body through some mysterious "fall", but seeks again to return to the One in an endless series of escapes from the body. Descartes as a Christian accepted this notion of innate ideas, but attributed the ideas to the Creator as "the mark of the workman impressed upon his work".[43] Moreover, while Plato had thought of these ideas as participating in the Absolute in an objective way, Descartes gave them an idealist significance by supposing that they were simply objects of our consciousness that as such (with the sole exception of the idea of God) give us no certitude about extra-mental reference. Modern philosophy has struggled ever since to get out of this "solipsism" or enclosed self-consciousness.

The point made by behaviorists in psychology is valid, in that "As a thing acts so it is". Thomas Aquinas, accordingly, sought to establish the uniquely spiritual character of human behavior marked by intelligence and free will by five empiri-

---

[42] See the text from Plato's *Meno*, with the figures drawn to illustrate the Socratic argument that "learning is remembering", in Deely 2001: 45–63, "The Socratic Method".

[43] Descartes 1641: *Meditations on First Philosophy*, Meditation 3, "Of God: that He exists", p. 142.

cal, behavioristic arguments.[44] In our intellectual culture today, practitioners of modern science, still hampered by the Enlightenment reduction of science to its ideoscopic dimensions, do not like to admit the validity of these arguments, whence modern scientists have often tried to reduce the difference between animals and humans to a matter of degree. But the actual development of modern science has in fact greatly supported the cenoscopic arguments earlier demonstrating that the transition to human animals in the course of evolution is like pregnancy in a woman: a matter of "either/or", "yes or no",[45] as we will next try briefly to show.

The first of Aquinas' arguments is that animals act deterministically by estimations wholly tied to the sensible, while humans exhibit free will evidenced by the way they choose among alternative behaviors according to a notion of good which transcends direct sensible instantiation, i.e., according to criteria which are not reducible to sensory objects. Modern scientific cultural anthropology has enlarged the evidence for this argument by showing that humans have elaborately different *cultures* and at least 7,000 living *languages*[46] that are not determined by "instinct", i.e., the sensory estimation animals make of objects in relation to themselves as desirable, undesirable, or safe to ignore.[47] While undoubtedly humans like other animals tend to be socialized by imitation and group

---

[44] Aquinas 1259/65: *Contra Gentiles* II, ch. 66.

[45] See the form of this argument directly based on the work of modern genetics, "The Emergence of Man: an enquiry into the role of natural selection in the making of man", in Deely 2009: *Realism for the 21st Century*, Reading 1, pp. 21–53.

[46] See Raymond G. Gordon, Jr., ed., 2005: *Ethnologue: Languages of the World*, where 7,299 languages are listed.

[47] While there remain experts who question this, none have proved the contrary. In addition to the decisive investigations of Sebeok cited above (see note 26 in this Chapter, p. 78), see Christophe Boesch and Michael Tomasello 1998: "Chimpanzee and Human Cultures" (with peer commentary and authors' reply).

and environmental pressures, the languages, technologies, arts, and religion of each culture are highly creative and contain points of originality that cannot be explained simply as environmental adaptations. Enculturation — *pace* Radcliffe Brown and his confreres[48] — does not reduce to socialization. They differ by true inventions and an ability to learn that jumps over dead generations to learn *directly* even from the thought of individuals no longer encounterable in social life, not by mere serendipity, which the dictionary defines as "an aptitude for making desirable discoveries by accident".[49] Can we attribute Einstein's theories of relativity, or the invention of the computer, to environmental adaptation or mere accident?

Aquinas' second argument is that while brute and human animal senses alike have special concrete objects because they are acts of bodily organs each of which has a specific object, human intelligence alone is of universals, i.e., not tied to the object as sensibly instantiable. For example, one can think the number 5, while my eyes see only five black marks on a sheet

---

[48] "The complete reduction of culture to social system, as characteristic of British anthropology and of American anthropology since the time of Radcliffe-Brown (1881–1955), is an oversimplification. Earlier American anthropology, with its concept of culture as 'superorganic' and wholly transcendent to social system, went too far in assigning autonomy to the cultural vis-à-vis the social. But the opposite view, that culture is nothing but the uniquely complex forms of social organization proper to man, also goes too far" (Deely 1980: "The Nonverbal Inlay in Linguistic Communication", note 7, pp. 215–216; further developed in Deely 1982 Part II: "Language, Knowledge, Experience", pp. 87–123.

[49] Noam Chomsky 1966: *Cartesian Linguistics: A Chapter in the History of Rationalist Thought*, claims that there is a common "deep structure" common to all languages; but this hardly amounts to more than saying they are all to a degree logical. Stephen Pinker 1994: *The Language Instinct*, claims that human communication is an evolutionary product that has rendered the brain much like a computer, an "artificial intelligence", yet he strongly distinguishes human from animal languages. George P. Murdock and Douglas R. White (1969, 2008) code 186 markedly distinct cultures that provide a world-wide sample of some 1,250 sub-cultures.

of yellow paper. It has been shown that primates can count up to 9 sensibles, learning by trial and error. A series of 35 images differing in color, size and position were used in teaching a chimpanzee that was rewarded with a banana if it touched four pictures; but if it made a mistake, it got nothing. These animals could also recognize that, for example, 7 instantiated sensibly is greater than 6 likewise instantiated. One of the researchers concluded that these chimps "share with humans the ability to master simple arithmetic on at least the level of a 2-year-old child".[50] Similar learning has been observed in birds, for example, Otto Koehler[51] showed a raven objects in sets of 2 to 6 and it learned to gain a food reward by matching these objects to black dots from 2 to 6.

Aquinas' third argument centers precisely on the difference between human language and animal modes of communication.[52] As already noted, animal communication is explicable by sensory estimations and consists mainly in calls that warn of approaching danger, or summons to a mate or an offspring, while humans invent new languages that not only warn or summon or at most name an object of the senses, but (as Aquinas points out) refer to such concepts as "truth" or "relation" or "syllogism" — concepts that do not refer to anything

---

[50] See Beth Azar 1999: "Researchers counting on animals for clues to math". Given the naïvety of this assessment, it is perhaps ironic that the original page for this <www.apa.org/monitor/apr99/math.html> has been removed, and the reference remains available rather only at "Science News for Kids" <http://www.sciencenewsforkids.org/articles/20031008/refs.asp>.

[51] Reported in Chapter 12 of Bernd Heinrich 1999: *The Mind of the Raven*. Cf. <http://www.angelfire.com/id/ravensknowledge/birdbrained.html>.

[52] See the 1914 classical work of the pioneer in animal psychology, Jean-Henri Fabre (1823-1915), entitled *The Mason Bees*. Of course, the rigidity of insect estimative power becomes progressively loosened in higher animals, but even in the secondary or "internal" senses of humans remains to a degree fixed. Also see Lauren Kosseff 2011: "Primate Use of Language"; Sebeok and Umiker-Sebeok, eds. 1980: *Speaking of Apes: A critical anthology of two-way communicaion with mani*; Stephen Hart 2010: "The Animal Communication Project" at <http://acp.eugraph.com/>.

material and sensible.[53] Linguists have sharpened this argument by showing that true language contains *syntactical* words, such as "the", "and", "therefore", and *negatives* such as "non-being", "nowhere", that stand for objects terminating purely mind-dependent relations (in contrast to mind-independent relations such as "parent and child").

In the 1930s, W. N. and L. A. Kellogg compared the development of a young chimpanzee they named Gua to that of their own newborn son, but concluded that Gua never used a true language. For some ten years after 1960, Allen and Beatrice Gardner studied a female chimpanzee, Washoe. She learned some 250 signs from American Sign Language, hand signals that deaf humans use. Although most of these were learned by repeated imitation of her handlers, sometimes Washoe imitated spontaneously and even combined signs, for example, "water" and "bird" to mean a swan. Without specific instruction Washoe could transfer signs; for example, she learned the word "more" when tickled, then spontaneously applied it to other actions or things. She could name pictures as well as objects. Washoe also used signs to care for a young chimp she adopted.

In 1966 Anne James and David Premack studied a six year old female chimpanzee, Sarah, using plastic chips, differing in colors and shapes, each representing an English word. She was given an apple to eat if she could identify the appropriate chip, and she learned to do this (hardly a surprise). When these chips were strung in a series, Sarah learned to make what her handlers[54] considered to be "sentences", such as "Mary-to-give-apple", with as many as eight units and using 130 signs. Yet Sarah never asked questions.[55] In 1972 Francine Patterson

---

[53] For details of the texts in Aquinas, see Deely 1971: "Animal Intelligence and Concept-Formation".

[54] In the standard "Clever Hans" fashion, as Sebeok put it, of seeing in the destination what was actually coming from the source!

[55] I.e., what the handlers construed as questions, for one can be sure that

also taught the gorilla Koko some hand signals from the American Sign Language code. Duane Rumbaugh and Sue Savage-Rumbaugh taught chimps and bonobos, notably a female bonobo, Kanzi, to use a computer keyboard with a design on each key corresponding to some object or behavioral pattern. When Kanzi struck a meaningful combination of signs she was rewarded (thus learning what her handlers considered "meaningful" — a question of who was training who). She also combined keyboard signs with gestures, and could respond correctly — i.e., behaviorally — to 70% of new commands given by a hidden speaker. To the command "Give the dog a shot", Kanzi responded by injecting a toy dog with a syringe. Recent studies also show that chimps can form "categories" represented by a single sign, or by combining signs; for example, one chimp called a watermelon "drink-fruit". Herbert Terrace taught a chimp called Nim to use signs such as "angry" and "bite" to express anger. Yet Nim never put together any sets of signs but only imitated sets created by his human teachers.[56]

Thus linguists point out that these behaviors reduce to signals of warning or attraction, and convey very limited information, such as, in the case of bees, the direction and distance to fly to find the right flowers. We intelligent persons may see that this conveyed information is true, but there is no reason to conclude that the bee knows the difference between truth and falsity, and it is very easy for us to deceive a bee by a tricky signal. In spite of the belief of some of these researchers that they had taught their animals true language, this research of many

---

Sarah at times would make clear that she wanted something or other, which is the behavioral equivalent of "Where is ... ?" linguistically expressed.

[56] This difference is what Poinsot 1632: *Treatise on Signs* (see esp. Book I, Question s, and Book II, Question 5) identifies as the difference between signs based on stipulation (*signa ad placita*) in species-specifically human linguistic communication and signs based on custom (*signa ex consuetudine*) which are generically shared or sharable by humans and other animals in social communication.

years has generally convinced psycholinguists and semanticists to the contrary.

True languages, i.e., species-specifically human systems of linguistic communication, however, such as humans have invented in many different forms can, in all its forms, be used to make verifiable statements, both true and false. When we look for the reason that this is possible, as semanticists have done, we find that animal signals lack *syntactical* relations such as are signified by such terms as "and", "therefore", "predicate", etc. — terms that do not signify anything sensible in the physical world, but only the mind-dependent relations that we form intellectually.[57] A statement that I repeat can be true or false, but only if I know its relation to the facts in the difference of that relation from the facts can I meaningfully assert that it *is* true or false; and this relation of the conformity of the statement to the facts is a syntactical relation that exists purely objectively, that is,[58] only in the consciousness of the one making the statement and the consciousness of his intelligent hearer who also knows the language in which he makes it.[59] Thus a dog may run to greet his master with a happy bark, because it has developed affection for its master, conditioned so by the treatment it has received from the master over time, including feeding, petting, etc.; and the master may recognize that the bark is appropriate, yet this does not show that the bark is a linguistic assertion of a truth, because there is no evidence that the dog knows the logical relation between his bark and the master's identity, but only evidence that the dog is pleased to have its master present.

Therefore, it must be concluded that these behavioral observations demonstrate that while other than human animals

[57] On this see Deely 2002: *What Distinguishes Human Understanding?*

[58] See Deely 2009a: *Purely Objective Reality.*

[59] Thus the famous observation of Maritain 1957: 55 that all "animals make use of signs without perceiving the relation of signification". See the YouTube presentation of "A Sign is *What?*" at <http://www.youtube.com/view_play_list?p=E9651802BCDC14BF>

communicate by signs, there is no evidence that they can communicate *truth*, since that requires syntactical language, a specifically linguistic communication. Aquinas gives this as one of five arguments, all of which are behavioristic, to show that humans have a kind of intelligence that is "abstractive" in the sense of not reducing to the perception of material objects, since all material things are quantitative and singular, in contrast to relations which extend to many terms and thus, when recognized as such, these relations makes that awareness an awareness of something universal, over and above the particulars related.

The fourth argument of Aquinas — a very powerful one — is that our senses, although they tell us that we are sensing what we are sensing, that is, are *reflexive*, are only imperfectly so. The higher sense is aware of what the lower sense detects. I cannot see the back of my neck, because material things are quantitative and thus have parts outside parts. Material things have as their first property quantity, defined as "having parts that have only common boundaries", and hence a quantified object cannot have all its points present to each other. Hence the senses, restricted to awareness of objects as material, cannot be the source of true self-consciousness, because to be self-conscious is to be present to one's self as performing an intellectual act at the very same level that the awareness proper to the act is located, that is, self-aware and not only other-aware.[60]

---

[60] Note that the totality of reflective self-consciousness does not mean "totally self-comprehensive", since our human knowledge of objects is never perfectly exhaustive, but only that in intellectual knowledge the object of which we become reflexively aware is the very awareness itself presenting an object intellectually, in contrast with the higher senses which are aware of the objects of lower sense but not of that object as their own awareness. Descartes was wrong to think that he was present to himself as a "clear and distinct idea". There are many degrees of self-consciousness, and in this life they are all very imperfect because of the dependence of the intelligence on the senses; and only at the intellectual level does the awareness itself become reflexively objectified.

In the continuing Mind-Body controversy there is often much confusion due to a failure to inquire whether the term "consciousness" is used to mean that kind of consciousness common to all animals or that further kind species-specific to human animals.[61] Anyone with an animal pet knows that Descartes was absurdly wrong in claiming they are automatons without some kind of consciousness. But is it not equally absurd to suppose that "consciousness" does not have levels? Do we not observe a vast difference in the sort of consciousness required while driving to stop at a stop-light and our consciousness of how to answer the arguments that a companion in the car is making against what we have just said? Even if I try to process my primary sense data by using the images provided by my secondary senses of memory, imagination, etc., I cannot at the same time sense an image and sense myself sensing that image.

By contrast, *self-consciousness* is precisely the ability intellectually — at the level of understanding[62] — to know myself knowing something, that is, I know that I know that I know, etc. This was the element of truth in Descartes' *Cogito ergo sum*; namely, that human persons are self-conscious; but this does not mean, as Descartes seemed to think, that we can have a pure self-consciousness independent of our primary and secondary senses. If you try just to think of yourself thinking you will find yourself thinking of nothing as some practitioners of "centering meditation" discover and suppose this to be a mystical experience. Human thought is not itself its own direct object. To know that I think I must first think about something. Even if I think about my thinking, that thinking thought about must be about something other than thinking.

---

[61] For a rather full discussion classifying the various positions of this topic (but with the *caveat* that the entry is drawn squarely from the late modern standpoint of so-called analytic philosophy), see <http://en.wikipedia.org/wiki/Philosophy_of_mind>.

[62] See Deely 2002 and 1994: *What Distinguishes Human Understanding?* and *The Human Use of Signs*, respectively.

Human intelligence in this bodily life, therefore, is not perfectly transparent to itself in such a way that I can know the essence of my soul simply by introspection as a pure spirit. In so far as I know who I am it must be done in the same way I would know anything else, by seeing how I behave and gradually sorting out my core identity from the tangle of irrelevant and superficial aspects of myself. What introspection can reveal that is not available to others is what goes on in my internal imagination and other sense data processing skills. But that is not intellectual activity as such. Thus, our self-consciousness is always background awareness, unless and until we make an effort to objectify it in its own right, to bring it into the foreground of awareness; and then the content we find is the essential content discovered in our sense data, but now presented within a reflexive *self-awareness* of the very act of understanding itself as itself having that content to consider.[63]

Yet before we can perform effective intellectual acts, we must have sense data that has been well-processed by the secondary senses whose organ is the brain. This processing can be partly unconscious, as is well illustrated by studies that have been made on "creative thinking". A famous example is the discovery by the chemist August Kekulé (1829–1896) of what is called the "benzene ring", an important feature of some molecules.[64] He had for many days tried without success to figure out what the chemical data seemed to be saying. Then one morning as Kekulé was waking up and was still half-dreaming, he seemed to see a snake run across the floor and then suddenly seize its tail in its mouth and roll out of the room! At

---

[63] The error of what is called "Transcendental Thomism" is to attempt to reconcile the epistemology of Immanuel Kant with that of St. Thomas by making our knowledge of a thinking, willing subject prior to our knowledge of sensible things. Quite clearly to the contrary is Aquinas 1266: *Summa theologiae* 1, q. 87, art. 3. For an analysis of this text in detail, see Deely 2008: 21–26.

[64] See H. C. Von Baeyer 1989: "A Dream Come True". See also John Lienhard 1989: "Inventing Benzene".

that moment he had the insight that the atomic units in the molecule form a circle.[65] Evidently, even when he was asleep, his mind had been somehow working the while, probing his secondary senses for an appropriate image that finally emerged as the snake-dream which supplied an analogy or model for conceiving intellectually the benzene ring.

Such stories are told about many discoveries, and what they show is that an inventor works a long time on a problem without the intellectual insight that would solve it until, often suddenly and during a time when maybe the inventor is thinking of something else, the right *image* pops up that makes possible his intellectual insight that answers his problem. This process is the basis of the technique of "brainstorming",[66] in which a group of persons trying to solve a problem keep proposing ideas without criticism, even ideas that seem wild and foolish, in the hopes that eventually someone will really see the answer. This shows that the sense data processing powers have a certain independence of the intellect, and can continue to work through chains of association, analogies, and comparisons of images even when this is not at the focus of our attention.

Aquinas' argument to show that the human consciousness, unlike that of animals, is capable of *perfect reflexion* and hence non-material is supported by the famous theorems of Kurt Gödel,[67] according to which it is impossible to prove that any logically formal system is complete or consistent without adding further axioms in the terms of that system. Aquinas argues that through our senses we can consciously know material objects, but only through our non-material intelligence can we know that we ourselves are existent beings who are in the

---

[65] See the mentions of Kekulé in Robert Weisberg 2006: *Creativity, Beyond the Myth of Genius*.

[66] See "Brainstorming" at <http://en.wikipedia.org/wiki/Brainstorming>.

[67] See Lucas 1961: "Minds, Machines and Gödel", and 1970: *The Freedom of the Will*.

process of knowing these objects; and we can know that we know that we know ... *ad infinitum*.

The fifth and last argument of Aquinas is based on the fact that a sense organ can be injured by too intense action of sensible objects on them (for example, we can be deafened by too loud sounds). The clearer our abstract thinking the stronger our intellects become, since one who knows the greater is able afterwards to know the lesser. This argument is related to Aquinas' discussion of angelic intelligence, in which the higher the angel is in the celestial hierarchy the fewer, not more, ideas the angel has, because each idea is of such great comprehension.[68] This argument is supported by the observation that the work of great geniuses often flows from some one insight they acquire early in their career.[69]

---

[68] Aquinas 1266: *Summa theologiae* I, q. 55, art. 3.

[69] In Aquinas' own case, it is often claimed that his original intuition, evident in his first philosophical work *De Ente et Essentia*, is the "real distinction of essence and existence". The prevalence of this inaccurate claim — which Ken Schmitz nonetheless (2010: 121; and fairly enough, given the profound development the distinction has received in Neothomistic hands) would not go so far as to call an outright fairytale — is rather mystifying, given that Aquinas himself expressly states that the distinction between essence and existence is a matter self-evident to all — otherwise, why would we worry that something might have happened to a friend who missed an appointment?! Aquinas cites examples of the distinction as he understands it in the work of Boethius and Aristotle, thus from the earliest times of both Greek and Latin philosophy. The actual position of Aquinas has been expressly outlined in Deely 2001: 290–296, "A Note on the Distinction between Essence and Existence". What was actually original in Aquinas concerning the essence/existence distinction was not the distinction itself, but the focus upon existence as an effect requiring a direct cause, rather than simply presupposed as in Aristotle's own analysis of causation. Considered as an effect in its own right, analysis of existence quickly shows its cause can only be a being in which essence and existence are *not* distinct, which in turn leads to Aquinas' doctrine of creation as the here and now dependency of finite being upon a purely actual source, in contrast to the (still) common (mis)understanding of the creation of the universe as a "past" event.

In our opinion, more fundamental to Aquinas' thought than his identification of the essence/existence distinction as a truth self-evident to all was his profound recognition that the whole of natural human knowledge

Therefore, human intellection is not what logicians call a "formal system" of logical deduction from a small set of arbitrary axioms. The mathematician David Hilbert (1862–1943) in the 20th century proposed the model of such a formal system and thought that mathematics could be reduced to it, while Bertrand Russell (1872–1970) proposed the idea that mathematics itself could be reduced to pure logic.[70] In such attempts the basic error is to reduce logic to the manipulation of symbols that stand for sets of objects defined only quantitatively, and not also by their *intention* or reference to specific kinds of things that we know from experience really exist. For a time, the school of Logical Positivism that originated with the so called "Vienna Circle", of which Rudolf Carnap (1891-1970) was a leader, thought that all of objective knowledge, which they held was limited to empirical science, could be articulated as such a formal system.[71] These philosophers thought that such a formal system would eliminate all ambiguity of language and at last realize the Cartesian ideal of "clear and

---

rests on the senses as a direct awareness of mind-independent aspects of our physical environs, a point in Aristotle not clearly seen by earlier readings of the Stagirite, not even by Aquinas' mentor St. Albertus Magnus; for it is precisely on this point that the overcoming of idealism in the theory of knowledge depends, as many of the great commentators on St. Thomas have expressly pointed out, such as: (1) Joseph Gredt 1924: iv — "the integral natural realism of Thomistic philosophy, the doctrinal heart of which consists in excluding from the intuitive cognition of external sense any least trace of a mental representation, is ... the one and only way of avoiding idealism"; (2) Jacques Maritain 1959: *The Degrees of Knowledge*, p. 118 note 1, which makes the same point; and (3) John Poinsot who made the point even before modern philosophy had gone the full way of idealism, 1632: *Treatise on Signs*, Book III, Question 2, 312/3–6. On this last text, see the historical contextualization of the point in Deely 2009b: *Augustine & Poinsot. The protosemiotic development*, esp. Section 12.8. "Poinsot 1632 on the point Hume 1748 deemed beyond consideration", 153–156.

[70] See A. D. Irvine 2010: "Bertrand Russell", and 2009: "Russell's Paradox".

[71] See Mario Murzi 2001: "Rudolf Carnap (1891–1970)", and 2004: "Vienna Circle".

distinct ideas". Gödel's Theorem, mentioned above, radically eliminated Logical Positivism.

To sum up, therefore, the facts of human and animal behavior show that animals and human persons have much in common, yet they also differ remarkably in that the human brain has so evolved that it makes possible an ensoulment by a spiritual soul directly produced *ex nihilo* by the First Cause, as is every being of the finite order, yet wholly within the order of nature even though exceeding the potency of matter, as we have noted from Poinsot's commentary on Aquinas.[72] This spiritual soul has both the physical powers required for the unified function of its body and for the acquisition of information from its bodily primary (or external) and secondary (or internal) senses, as well as from spiritual powers of intellection and will to analyze and freely use information acquired to satisfy human needs. This complex places the human person on the borderline between the physical and spiritual regions of our universe, and requires study first by comparative psychology (that is part of natural science) and second by personal psychology (that pertains to metaphysics, but which works by analogy from the results of comparative psychology in the social order).

Thus modern psychology, in its unsuccessful attempts to reduce human consciousness to a mere difference of degree from animal consciousness, has (in spite of itself, as it were) clarified two points. First, human thought does not arise from innate ideas, but is dependent on the senses, primary and secondary, and hence on the wonderful complexity of the human brain. Continued research on the brain, therefore, will supply us with better instruments of thought and better ways of healing and using them. Second, science itself depends on a human intelligence that is immaterial, and hence able to transcend animal thinking, but only through a good use of animal thinking as its instrument.

---

[72] Recall Chapter 2, pp. 55–56 in note 27.

Human Affectivity and Freedom

These conclusions apply also to the *affective* as well as to the cognitive aspect of animal and human psychology that is today often confused by the term "feelings". For Aquinas, our appetites or drives (passions, affects) are, as such, not conscious. They are stimulated by images coming in animals from the secondary sense,[73] the *vis aestimativa* and in humans from the equivalent *vis cogitativa* ("discursive sense" or "particular reason", but which we would prefer to call the *evaluative* sense or even *estimative* power). This sense adds to an image a positive (attractive) or negative (repulsive) note, indicating the relation of that image to the well-being of the animal, human or not. This stimulation of an affective drive causes a reaction in the body that then becomes conscious as a change in bodily state, e.g., the positive image of food causes salivation in the mouth and "feelings" of hunger in the stomach, etc.

Therefore, just as the theoretical intelligence cannot form an abstract concept without an appropriate image ("phantasm", in the generic usage of the Latins), so also we cannot make a free, practical decision without an appropriate image, in this case coming from the evaluative sense. Modern neurology

---

[73] The term "cathexis" is useful in this regard, particularly in the usage proposed by Parsons and Shils 1951: *Toward a General Theory of Action*, where cathexis is used to designate the affective component that accompanies every cognition in the world of animals, human or not. The term can thus be used (e.g., in Deely 2009a: *passim*) to designate the fundamental distinction of psychological states into *cognitive* and *affective*, a division that thus has the same sense that Aquinas attaches to the division of purely objective being into *negations* and *relations*, that is to say, as constituting a division which is exhaustive and exclusive. Thus (Deely 2007a: 162) "the intraorganismic factors involved in the presentation and organization objectively of the content of the organism's experience are not only cognitive. To every thought concomitates a *feeling* that is equally intentional respecting the object signified, the content 'upon which' the thought is directed, and this feeling, or 'cathectic factor', is no less relevant to the animal's world than are the cognitive elements or 'ideas'," concepts in the generic sense of *species expressae*, in Aquinas' terminology.

is gradually locating all these powers of the human body instrumental to our spiritual powers, especially of the secondary
senses, in the human brain and the secretions of the hormonal
system that produce the changes of the body that we call "feelings". Psychotherapy is also applying this view to healing mental ills, both by drugs and by correcting bad conditioning of
mental habits.

Jesus as the Perfect Human

Therefore, theology can be enriched in its understanding of
moral life, of which Jesus is the exemplar, by these advances in
understanding in human psychology, both in its physical and
its spiritual character. During much of the twentieth century
the psychoanalytical theory of Sigmund Freud, with its pan-
sexualism (reduction of both intellectual and appetitive life
to the sexual drive and to an "unconscious mind"), was used
to discredit Christian moral tradition. Today Freud himself
is to some extent discredited by more critical psychological
research.[74] Thus, when the Gospel speaks of Jesus' "compassion" and other human feelings we can make use of better,
post-Freudian psychology to understand them and relate them
by analogy to Jesus' human spiritual sympathy, his human
intellectual freely chosen love, his elevation by grace (of the
"Sacred Heart"), and finally to his Divine Love, which he is in
his Divine Person in the Trinity.

Does science cast any light on Jesus' conception in Mary's
womb by which he became the perfect human? The Holy

---

[74] On Freud and religion, see David Bakan 1958: *Sigmund Freud and the
Jewish Mystical Tradition*, and Paul Vitz 1988: *Sigmund Freud's Christian Unconscious*. Freud's view of the human person was materialist and deeply pes-
simistic, as is evident in his two famous books *Totem and Taboo* (1918) and
Civilization and Its Discontents (1930). In the former work, Freud attributed
religion and religious ethics to a defensive denial of our hatred of our father
as a rival for our mother's sexual favors. In the latter, he argued that civilized
society can only exist by the painful, and never quite successful, suppression
of basic human instincts.

Office, now the Congregation for the Doctrine of the Faith, in 1960 cautioned against unseemly speculations on this subject.[75] What we know for certain by Christian faith is that neither in Jesus' conception, nor birth, nor after his birth, was Mary's virginity in any way diminished. This is complementary to her own Immaculate Conception and to her bodily Assumption at death, by which she is the new Eve, the only other perfect human than the Divine Son she virginally conceived. This is one of the points on which the *Qur'an* of Islam suffers from an obscurity. It refers several times to Mary, Mother of Jesus,[76] even to the point of seeming to recognize her Immaculate Conception. Yet the *Qu'ran* also hesitates to praise Mary, lest her eminent holiness derogate from the absolute Oneness of God, Allah (which, in Christian belief and cold logic, it does not). Rather, God's absolute Unity, by analogy to human self-

---

[75] See *Ephemerides Mariologicae* 11 (1961) 137–138. This was not officially published, but sent privately to certain bishops. Cf. René Laurentin 1991: *A Short Treatise on the Virgin Mary*, pp. 328–329; also the reply to recent critics who question this, by Msgr. Arthur B. Calkins 2005: "The *Virginitas in Partu* Revisited". Cf. Most 2010.

[76] The particularly relevant texts of the *Holy Qur'an* are, in the M. Pickthal translation, the following. *002.087*: "We gave unto Jesus, son of Mary, clear proofs (of Allah's sovereignty), and We supported him with the Holy Spirit. Is it ever so, that, when there cometh unto you a messenger (from Allah) with that which ye yourselves desire not, ye grow arrogant, and some ye disbelieve and some ye slay?" *003.047*: "She said: My Lord! How can I have a child when no mortal hath touched me? He said: So (it will be). Allah createth what He will. If He decreeth a thing, He saith unto it only: Be! and it is." *005.116*: "And when Allah saith: O Jesus, son of Mary! Didst thou say unto mankind: Take me and my mother for two gods beside Allah? he saith: Be glorified! It was not mine to utter that to which I had no right. If I used to say it, then Thou knewest it. Thou knowest what is in my mind, and I know not what is in Thy Mind. Lo! Thou, only Thou, art the Knower of Things Hidden?" *019.035*: "It befitteth not (the Majesty of) Allah that He should take unto Himself a son. Glory be to Him! When He decreeth a thing, He saith unto it only: Be! and it is. And Mary, daughter of 'Imran, whose body was chaste, therefore We breathed therein something of Our Spirit. And she put faith in the words of her Lord and His scriptures, and was of the obedient."

consciousness, is guaranteed precisely by the procession of the Son from the Father, as our self-knowledge is guaranteed by our self-knowledge. The same follows for the human will, by which we will our true end in analogy to the Holy Spirit.

Thomas Aquinas supposed that, naturally, the human soul is not infused until at least a month after insemination,[77] but in the case of Jesus there was no such period of "delayed hominization". He was created human soul and body together from the first moment in Mary's virgin womb, and at that very same moment already enjoyed intellectual self-awareness,[78] although for Aquinas in ordinary children this self-awareness naturally took place continuously only at about seven years of age.[79]

Current theology tends to pass over this question in silence.[80] Because according to Catholic doctrine (and according to reason) the human soul is spiritual, it has to be infused into rather than educed from matter, directly by a free act of God, while the human body is produced biologically by its human parents. Yet, because God is First Cause, he is also as regards existence the ultimate cause of the human body as well, albeit

---

[77] Aquinas argued that the material dispositions which call for the infusion of the spiritual soul take time to develop, so the fertilized egg begins with a vegetative soul but under a formal finality which develops the fetus to the point or "instant" at which an animal soul is educed replacing the vegetative, and then again through further development the dispositions are achieved which call for the infusion of the species-specifically human soul to animate the fetal body which becomes fully human only at that moment of infusion. Modern genetics, which has determined that the genotype is fixed at the moment of fertilization and remains as the one biological constant from conception to death, gives strong grounds for arguing that Aquinas' view was mistaken on medical grounds; and this is the position taken in Ashley 2006a: "When Does a Human Person Begin to Exist?", Chapter 20 of *The Ashley Reader*, pp. 329–368. Deely agrees more with Maritain 1967, esp. Section 6.

[78] Aquinas 1266: *Summa theologiae* III, q. 34, art. 4. See further Ashley 2006a: "Christology from Above: Jesus' Human Knowledge According to the Fourth Gospel", Chapter 9 of *The Ashley Reader*, pp. 109–124.

[79] Aquinas *Summa Theologiae* III "Supplementum", q. 43, art. 21.

[80] See again the remarks of Msgr. Calkins 2005. See also Calkins 2006.

using the parents as his instruments in the order of secondary causes (the order of interactions among finite beings). Thus human conception, soul *and* body, is ultimately the act of God (as is the whole of creation), preparing a suitable human body through instrumentality of the conjugal relation of the parents and completing it as a single substance by the direct creation of its soul.[81] Even in today's use of artificial conception in a test-tube, God creates the soul when the material body is proximately prepared, although this deprives the child of its right to be bonded to its parents by their committed and loving marital act.

In natural conception, the ovum and sperm are originally cells of the respective parent's bodies which, when mature, are separated from their bodies but continue to act as their procreative instruments. Each is haploid, that is, contains only half the genes required for a new human body, to which, therefore, the male and female parent contribute equally. The male contribution, however, can contain either an X chromosome, or a Y chromosome necessary for a male child, while the female parent contributes only X chromosomes. Moreover, it is the male sperm which is the more active agent giving to the female ovum the added energy required for the embryo to begin to develop to the point that it implants in the mother's womb and begins to draw nourishment from her.

The female ovum, up until implantation, provides the greater part of the extranuclear material of the conceptus. In the nucleus of the conceptus the two sets of genes are fused to produce the new body of the child, and when this comes into existence the Creator, First Cause of the entire process, directly creates the soul as infusing the body, as its one substantial form. Of course the term "soul" is not used in current science; but in Chapter 4 we will see that without this concept accurately defined the current dilemma in science about the Mind-Body relation cannot be solved.

---

[81] See *Catechism of the Catholic Church*, ## 363, 366, 1793.

Thus, in the case of Jesus' conception, Mary must have produced a mature ovum with its haploid genes, that would eventually be menstrually discharged, as normally with virgin women, but with her consent to God's plan God created in it a soul, supplying the necessary complement of her haploid genes but with a Y chromosome and the motor energy naturally supplied by a sperm. This would of course be a miraculous conception, but would otherwise be perfectly normal. Yet, beyond the miracle of a virgin generation and birth of a human child, the child in this case would be the incarnation of the Son of God, because the existence of this child would not be that of a human substance but of the divine super-substance of the Son of God, Second Person of the Holy Trinity. By the notion of "substance" is meant a natural thing that has independent existence, "being in itself". In a natural human conception, the independent existence given to the child's nature is not more than a natural existence dependent, like all natural things and the entire universe, on the First Cause. By contrast, in Jesus' incarnational conception, the existence given to his body and soul was the divine, eternal existence of the Son of God, member of the Holy Trinity which is the One God, the I AM, Existence Itself, the Uncaused Cause. Thus, Jesus is one Divine Person who has both a divine and a human nature. He is, however only one person, not two, because the term "person" signifies an intelligent and free substance[82] precisely as existent.

Therefore, although it is certainly a mystery transcending human reason how the Divine Son could become also truly human, yet a scientific understanding of human conception makes clear that there is no *contradiction* in this article of Christian faith, and even that it is a most appropriate way by which

---

[82] The term "substance" can be applied, but only analogically, to a Divine Person as this Person has divine existence. The Son has existence from the Father, not as from an efficient cause but only through a subsistent relation. *Summa theologiae*, I, q. 29, art. 2, and q. 39.

God could make known to us his otherwise hidden presence and community with us, since it perfectly corresponds to our human way of understanding reality.

Recent attempts by some theologians to express the traditional formulation of the transcendent mystery of the Incarnation in ways more understandable to our culture have attempted to develop Christologies "from below", beginning with what the New Testament tells us about the humanity of Jesus. Unfortunately, this speculation has sometimes fallen far short of what the Catholic faith holds about the Incarnation. Christology must insist equally on the "from above" and the "from below" — that is, on the Divinity *and* Humanity of Jesus, because they are equally true, although it is through his Humanity that his Divinity has been made known to us, and it is from the analogy of his likeness to us that we must come to some understanding of his Divinity. To use modern anthropology, psychology, and sociology in a Christology from below to make the Incarnation intelligible to our culture, as Roger Haight, SJ, did in presenting Jesus simply as "Symbol of God", reduces the mystery of the Incarnation to empty rhetoric.[83] On the other hand, to refer to the scientific embryology of his virginal conception disturbs our scientistic culture as it needs to be disturbed, yet makes it intelligible to those who realize, as here argued, that the consistency of modern science depends on the acknowledgment of a First Cause of existence as the basis of all causality.

Aquinas followed the Greek doctors in supposing that human ensoulment took place some weeks after conception, and this did not fit very well with his theological affirmation that the Incarnation took place at the moment of the virginal conception. By assuring us that the human individual genome is

---

[83] Vatican, Congregation for the Doctrine of the Faith, 2005: "Notification about the book *Jesus Symbol of God* by Fr. Roger Haight, SJ", online at <http://www.vatican.va/roman_curia/congregations/cfaith/documents/rc_con_cfaith_doc_20041213_notification-fr-haight_en.html>.

actually complete at conception, as noted above,[84] modern embryology has relieved theology of this embarrassment.

This approach, however, confronts us with question about the meanings of the term "person". Our culture emphasizes "personalism", "personality", "the equal dignity of persons" and "the rights of the human person", yet finds it difficult to say precisely *what* a "person" is. Thus "pro-choice-ers" argue that the embryo is not a person, because it is only *developing* into a person. While referring to scientific embryology they yet contradict its research results, because such research clearly establishes that the embryo is already an independent substance identical in first act with its mature state. On the other hand, animal rights advocates, such as Peter Singer, claim that all animals have rights just as do human animals, because all animals can feel pain.[85] Others argue that the definition of

---

[84] See previous note 77.

[85] Singer, in a 2005 article, "The Great Ape Debate", is more precise, but fails to define "intelligence" <http://www.utilitarian.net/singer/by/200605--.htm>: "I founded the Great Ape Project together with Paola Cavalieri, an Italian philosopher and animal advocate, in 1993. Our aim was to grant some basic rights to the nonhuman great apes: life, liberty, and the prohibition of torture. The Project has proven controversial. Some opponents argue that, in extending rights beyond our own species, it goes too far, while others claim that, in limiting rights to the great apes, it does not go far enough. We reject the first criticism entirely. There is no sound moral reason why possession of basic rights should be limited to members of a particular species. If we were to meet intelligent, sympathetic extraterrestrials, would we deny them basic rights because they are not members of our own species? At a minimum, we should recognize basic rights in all beings who show intelligence and awareness (including some level of self-awareness) and who have emotional and social needs. We are more sympathetic to the second criticism. The Great Ape Project does not reject the idea of basic rights for other animals. It merely asserts that the case for such rights is strongest in respect to great apes. The work of researchers like Jane Goodall, Diane Fossey, Birute Galdikas, Frans de Waal, and many others amply demonstrates that the great apes are intelligent beings with strong emotions that in many ways resemble our own." Note that Singer's notion of "intelligence" is not only vague, but seems to beg the question of whether there are not only different degrees of intelligence between human and other animals but also a difference in kind — a rather serious issue to dodge. Cf. Adler 1967: *The*

"person" is a matter of social consensus, yet denounce sexism and racism.

For Catholic Faith and for Thomistic anthropology, however, a person is an independently existing thing (substance) that has the properties of intelligence and free will, and it is on the basis of this definition that moral rights and obligations must first be understood. This definition includes members of *homo sapiens sapiens*, extraterrestrials if they exist, pure spirits, and the Three Divine Persons, although these three orders of persons are such only by analogy. Human intelligence depends upon sensory stimuli and the human brain in the formation of concepts; angelic intelligence has no material dependence but forms concepts directly in response to the divine creative action sustaining in existence the interacting finite substances of the physical universe;[86] the Divine Persons, in sharp contrast, have no dependence upon concepts at all and are distinguished only by their relations within the One Godhead in which the whole of creation is simultaneously present in the divine awareness.

Thus persons, in these three analogical, related meanings of the term, are highest in the hierarchy of beings, as even the most materialist of scientists must admit with regard to human and (so far hypothetical) extraterrestrial persons. Scientists, by the fact that they are persons, are able to research the material

---

Difference of Man and the Difference It Makes, and Deely 2002: *What Distinguishes Human Understanding?*

[86] For St. Thomas, the divine creative action is to the intelligence of angels what the stimuli of sense are to the intelligence of humans, to wit, the basis upon which the intelligence proceeds to fashion its ideas interpretive of the stimuli. The angelic intelligence, thus, actively forms, not passively receives, its ideas or concepts. Thus the ideas of angels are not "innate", as is commonly bandied about, nor are they "infused" normally (although this can sometimes be the case); normally, angelic ideas are the product of the activity of the angel's own intellect and process of thinking in response to the creative action of God under which finite beings enter and exit within the interactions of secondary causality which make up the physical universe. See the extended treatment in Poinsot 1643: "Tractatus de Angelis"; English analysis of the Thomistic texts in Deely 2004: "The Semiosis of Angels".

universe and to a degree achieve control of it on the basis of the knowledge their researches yield, and hence transcend in that measure all the material things they study and control. Aristotle, because the main physical force he knew was heat (notably from the sun), and because the human heart was the hottest organ of the body, held that the heart was the primary organ of the human body and the direct instrument of human intelligence. Later science showed that it is the human brain that is the most energetic organ of the human person and guides intelligence, although the brain has itself been, along with the rest of the mature human body, constructed by primordial embryonic organs programmed by the genome already completed at conception. The human person, therefore, is the reality accessible to our direct observation, from which, by abduction and analogy, we can come to some understanding, inadequate as it may be, of the spiritual realm of our universe, and finally of its First Cause.

Of course this First Cause, since it is Pure Act, cannot be understood to be, as monists tend to think, the *material* cause of the universe (a view Aquinas uncharacteristically dismissed as "stultissimus" — most block-headed). Instead, this First Cause or God as Father and Creator is the constant sustaining cause of the universe,[87] and God the Son is its Word, or formal cause, the pattern after which the creation is formed. Some commentators on Aristotle's works understand him to make God the final, but not the creating cause, of the universe; but Aquinas, rightly we believe, did not read Aristotle in this way.[88] To

---

[87] The creative causality is often compared to and called an "efficient causality"; but though such an analogical usage is not without some merit, when we consider that every efficient cause *presupposes something* to act upon, whereas the act of creation *presupposes nothing*, that creation is *ex nihilo*, it becomes clear that the creative action of God is best regarded as *sui generis* rather than, perhaps, as an example of "efficient action".

[88] For an example see Barry Smith 2006, where one reads the following summary of *Metaphysics* XII (Lambda) as follows: "God could not impart motion as the first efficient cause, because to do so God would have to be in

suppose that God who is super-intelligent does not know what he has created and conserves, while we, whom God created, can do so, is absurd, even though the absurdity is a consequence Aristotle himself did not reach.

Although the Trinity is One God and acts on the universe as One, the divine causality of creation can be credited analogically to the Father, and its formal causality to the Son, as also its final causality can be attributed to the Third Person of the Trinity, the Holy Spirit. The fourth kind of causality, material causality, cannot be attributed to God in any direct way, but only to the physical realm of his Creation. Thus the historical movement of the created universe, both as it is known through reason and as it is known through revelation, is a process of redemption, transformation, and elevation to eternal union with its Maker. This fact helps us to understand the mystery of the Holy Spirit. As the Father is Power, the Son Word, so the Holy Spirit is the Love that moves the universe, mired through sin, back again to the Father through the Son. Thus at Pentecost Jesus, after earning our salvation and having returned to the Father, sent the Holy Spirit to his Church to guide it to the Father in the course of history. Hence the history of the universe climaxes in the Paschal Mystery of Jesus' Incarnation,

---

motion, and if God were in motion, then God would be moved and movable. Besides, there is no beginning to the process of eternal motion, no creation. What is implicit in Aristotle's argument is that the first heaven has intelligence, or soul, in order to love the unmoved mover and so allow the latter to function as final cause". Aquinas, in his comments *In Metaphysicam* XII, says rather that Aristotle seeks to show the perfection of the First Cause by showing that the action of the First Cause surpasses material agents and even the spiritual movers of the heavens because it is analogical to the way a goal (final cause) moves a free agent, which is a mode of causality superior to all others. Thus he by no means *denies* God is an efficient cause, but only argues that his agency is perfect: the creative action brings things into existence, as does an efficient cause, but it brings things into existence *from nothing* and sustains them *outside nothing*, whereas an efficient cause in the Aristotelian sense brings things into existence by acting upon materials already existing. Still, there is no doubt that, of the four causes of Aristotle, the creative action of God is *more like* the efficient cause than any other.

Crucifixion, Resurrection, Ascension, and Pentecost, and will be completed through the preaching of the Gospel in the Holy Spirit until the Last Judgment.

To conclude this chapter, let us observe simply that we have considered the ordering of the vast variety of specifically different bodies and spirits in the universe, especially the border-species, those strange spiritualized animals, human persons. Just because, as embodied persons, we stand at this border-line (and thus at the top edge of the proper scope of natural science), every truth that science discovers about our universe enriches our reflexive self-understanding of truth itself. Furthermore, this conclusion is central to Christian theology, because God the Son has been sent by the Father in the power of the Holy Spirit to share our life as the God-Man. How has the Holy Spirit brought this to pass?

Chapter 4

THE UNIVERSE'S EVOLUTION, FREEDOM, AND PLENITUDE REVEAL GOD THE HOLY SPIRIT

Does the Universe Evolve?

One of the most striking features of modern science is that it has replaced its exclusive interest in uniform natural laws with an historical emphasis in the Big Bang cosmology and evolutionary biology. Natural laws are based on regular repetitive changes, while history is a chronology of unique events, i.e., events with unrepeatable aspects, even though occurring within in a context of laws. Yet some scientists think the Big Bang will reverse itself in a Big Crunch. Many also suppose that biological evolution is taking place extraterrestrially on many planets. Moreover, some leading scientists who rely on the Multiverse hypothesis hold that perhaps other worlds have other natural laws of development than has our own. Yet all such hypotheses depend on conjectural analogies to our own history, since the events of history can no longer be directly observed. Eyewitnesses, photographs and recordings may seem to give us direct information, but still require interpretation. Thus, since history has to do with an increasingly remote past that scientists cannot experimentally control, this means that current science, in so far as it rests on historical grounds, involves irreducible elements of uncertainty, the more so the further back our guesses reach.[1]

---

[1] See, for example, the discussions in Nancy Atkinson 2008: "Are the

Yet also this recent emphasis on evolutionary development means that there is a convergence between the scientific narrative and the Jewish-Christian-Islamic religious narrative stretching from God's creation of the universe from nothing toward some kind of ultimate goal to which God is providentially guiding that creation. This favors the monotheistic religions (and especially Christianity, which provides a more detailed sacred history than do either Judaism or Islam), as against the monist religions of paganism, including those of the Greeks and Romans. Monism, to the contrary, generally supports the "Myth of the Eternal Return", while agnostic and secularist scientists, who distrust myths, prefer such tenuous hypotheses as Multiverse,[2] the Big Crunch, and ETs that lack empirical verification.

Nevertheless, although Pius XII was cheered by the Big Bang Theory,[3] Christians have generally been fearful of evolu-

---

Laws of Nature the Same Everywhere in the Universe?"; Nawal Mahmood 2011: "Laws of Nature May Not Be the Same Everywhere. Revolution in Physics?"; *The Economist*, 2 September 2010: "Ye cannae change the laws of physics — Or can you?".

[2] Michael D. Lemonick 2004: "Before the Big Bang", p. 36, reports that Cosmologists Paul Steinhardt and Neil Turok of Princeton University "have a radical idea that could wipe away these mysteries. They theorize that the cosmos was never compacted into a single point and did not spring forth in a violent instant. Instead, the universe as we know it is a small cross section of a much grander universe whose true magnitude is hidden in dimensions we cannot perceive. What we think of as the Big Bang, they contend, was the result of a collision between our three-dimensional world and another three-dimensional world less than the width of a proton away from ours — right next to us, and yet displaced in a way that renders it invisible. Moreover, they say, the Big Bang is just the latest in a cycle of cosmic collisions stretching infinitely into the past and into the future. Each collision creates the universe anew. The 13.7-billion-year history of our cosmos is just a moment in this endless expanse of time. The hidden dimensions and colliding worlds in the new model are an outgrowth of superstring theory, an increasingly popular concept in fundamental physics."

[3] Doug Linder 2004: "The Vatican's View of Evolution: The Story of Two Popes", reports that Pius XII, in a 1951 address to the Pontifical Academy of Sciences, said that "it would seem that present-day science,

tionary theory ever since the days of Darwin. This distrust has several bases.

The First Objection to Human Evolution

The first objection to human evolution, of course, is the excessively literal understanding of the *Genesis* account held by much of patristic tradition. For that reason, Aquinas and the medievals generally also favored this literal view; and today, to meet liberal attacks on Biblical inspiration, the literal view is now pushed to its extreme (just as was discussion of the relation between sun and earth in the 1600s) by modern fundamentalists, both Protestant and Catholic. Yet, as Galileo pointed out in his letter of 1615 to the Grand Duchess Christina (a letter written in vain for the Inquisitors of his day, but which has since come to be regarded as a fundamental guide in mediating scientific discovery with biblical interpretation),[4] already St. Augustine, greatest of the Western Church Fathers, founder of hermeneutic theory and initiator of the Latin development of semiotic consciousness in general, made clear by the 5th century that the literal or historical meaning of Sacred Scripture need not be understood as we would a modern newspaper account, but may be intended by the Divine and human authors to be an inspired yet symbolic or analogical narrative.

---

with one sweep back across the centuries, has succeeded in bearing witness to the august instant of the primordial *Fiat Lux* [Let there be Light], when along with matter, there burst forth from nothing a sea of light and radiation, and the elements split and churned and formed into millions of galaxies".

[4] Recall from Chapter 2 our "point of the Galileo Affair", note 10 pages 41–42, particularly the "error in principle" Maritain points out (1973: 210), namely, the error of holding the development of sciences of phenomena — science in the modern sense, ideoscopic science — to be subject *in their own development* to theology and the literal interpretation of Scripture ("against which St. Augustine and St. Thomas [followed by Galileo in his letter of 1615] forewarned us"; cf. also Chapter 3, p. 77).

Why suppose that the final human author of *Genesis* claimed to have been told by God that the world was created in six days, or that Eve was literally tempted by a talking serpent? The author in question had no human way of knowing the distant past, nor does he (or she?[5]) say that God supplied that information. He expressed what he knew from the information available in his own time, long years (to put it mildly) after these events occurred. Under the inspiration of the Holy Spirit nonetheless, the author of *Genesis* used a simple analogy between a week's labor by a Jewish craftsman (as Jesus, the New Adam, himself was to be) and God's free creation of the world, an act that is infinitely beyond our human understanding. Since God has no need "to rest from his work", this analogy seems explicitly indicated by the biblical writer when he says (Gn 2: 1–3):

> Thus the heavens and the earth and all their array were completed. Since on the seventh day God was finished with the work he had been doing, he rested on the seventh day from all the work he had undertaken. So God blessed the seventh day and made it holy, because on it he rested from all the work he had done in creation.

How then are we to understand the narrative of humankind's fall into original sin (Gn 3)? Since God made a world that was "very good", and nothing is more obvious than that the present world of the human author of *Genesis*, just like our world today, was far from "very good", its condition must be somehow the result of creaturely sin, and in this sense "fallen", as Pope Benedict XVI has explained:[6]

> Faith tells us that there are not two principles, one good and one evil, but there is only one single principle,

---

[5] Come to think of it.

[6] Pope Benedict XVI 2008: "The Apostle's teaching on the relation between Adam and Christ", par. 6. See the further citation of this text in note 71 below, pp. 156–157.

God the Creator, and this principle is good, only good, without a shadow of evil. And therefore, being too is not a mixture of good and evil; being as such is good ... evil is not equally primal.

The alternative is the Gnostic view that the world was created by an evil or at least a stupid God. That angels, creatures wiser (or at least of a knowledge more comprehensive than the understanding of human creatures) also sinned is symbolized by a serpent tempter. Wisdom was attributed to snakes because the ancients especially admired wise men as able to find solutions to difficult problems, and a snake is able to escape enemies by slithering suddenly into hidden crevices, and can likewise suddenly attack its enemies before they have any awareness of their danger.

To say that Satan tempted Eve, and then that Adam, lest he lose his beloved companion, agreed to this sin is only to say symbolically that human sin begins from the nurturing or affective part of our human nature, often more evident in women, and then blinds our aggressive caution, generally more evident in men. The narrative says plainly that *all* humanity, men and women, shares in this sin and Adam's guilt. Because Adam was head of his the human race as his family he had the greatest responsibility, and thus was the most guilty when he fell short in that responsibility.

Hence original sin can be understood as "original" not only in being the first of all sins but also "not as an act but as a state",[7]

---

[7] *Catechism of the Catholic Church*, ## 403–406. It is often claimed that St. Augustine "invented" the doctrine of original sin, but in fact he simply clarified a traditional understanding of the Biblical texts: "For since death came through a human being, the resurrection of the dead came also through a human being. For just as in Adam all die, so too in Christ shall all be brought to life" (I Cor 15: 21–22), and "In conclusion, just as through one transgression condemnation came upon all, so through one righteous act acquittal and life came to all. For just as through the disobedience of one person the many were made sinners, so through the obedience of one the many will be made righteous" (Rm 5:18–19). It is also claimed that these texts speak only

that is, as a defect of all human nature, originally created by God to be very good, indeed, but nonetheless created as a spiritual nature that is animal in origin and finite in understanding, subject to the cumulative effects of bad choices over time. Hence every human child from and by its conception inherits "original defects" — the inherent limitations of a bodily nature dependent upon successful interaction with its surroundings, both physical and social, to survive — that weaken our deepest tendency to seek God, rather than to seek ourselves as if we were gods. Thus "original" sin can also be taken as the whole historical accumulation of human selfishness that we see every day around us in our world. As such sin is not directly God's work but the work of God's free creatures, and an expression of their inherent limitations.[8] God has not willed the moral evil in the world, and has given to human beings along with their intelligence the power to overcome its *physical* evils, which God did will in his creation not for their harm's sake but for some greater good. Cats eat mice, but cats need food to be (as they are) more wonderful animals than mice.

Charles Darwin (1809–1882), to whom the modern theory of evolution is mainly attributed, was a brilliant naturalist, an observer of nature in its variety; but he fell away from

---

of physical death, but infant baptism was practiced by the Church from its beginning; see, in the *Catechism* #1252, "Baptism, Infants, and Salvation".

Nonetheless, it is worth noting that prior to Augustine, notably, perhaps, in the Greek patristic writings which were inaccessible to Augustine by reason of the linguistic barrier, the interpretation of *Genesis* is considerably more complex and less literal than it became for the Latin church in the wake of Augustine's influence. Eastern Orthodox Christianity, for example, retains this earlier diversity down to the present day in dealing with the interpretation of "original sin" as reported in *Genesis*. And it was Paul who advised (*Colossians* 3:22) that slaves obey their masters rather than rebel against slavery. On original sin as also a state, and not merely the first misuse of freedom possible only when the first human animal walked the earth, see esp. Schoonenberg 1965, 1967.

[8] Aquinas 1266: *Summa theologiae* I, qq. 82–83. Deely 1997: esp. 71–79. And Benedict XVI in note 71 p. 156 below.

the Protestant fundamentalist faith in which he was raised and became an agnostic. He was throughout much of his life subject to constant illnesses and depression,[9] and explained his agnosticism by his struggle with the aforementioned classical problem of evil. In a letter of May 22, 1860, to another famous naturalist, Asa Gray (1810–1888), Darwin wrote:[10]

> With respect to the theological view of the question: This is always painful to me. I am bewildered. I had no intention to write atheistically, but I own that I cannot see as plainly as others do, and as I should wish to do, evidence of design and beneficence on all sides of us. There seems to me too much misery in the world. I cannot persuade myself that a beneficent and omnipotent God would have designedly created the Ichneumonidae with the express intention of their feeding within the living bodies of caterpillars, or that a cat should play with mice .... On the other hand, I cannot anyhow be contented to view this wonderful universe, and especially the nature of man, and to conclude that everything is the result of brute force. I am inclined to look at everything as resulting from designed laws, with the details, whether good or bad, left to the working out of what we may call chance.

Darwin seems to have been unacquainted with the answer to this question given by Aquinas, who pointed out that *moral* evil and *physical* evil are two quite different things. God creates freely; however, what is *creatable* attains perfection only in a finite way, as an extrinsic and perforce limited imitation of the unlimited divine perfection. A material world, moreover, is of its very nature liable to physical evils, since one kind of body is made out of the potentiality of some other kind of body,

---

[9] See Jerry Berman 2004: "Was Charles Darwin Psychotic?"; and the *Wikipedia* entry "Charles Darwin's Health".

[10] Darwin 1860: letter 2814, from *The Correspondence of Charles Darwin* volume 8.

and thus subject to destruction. The production of water out of oxygen and hydrogen destroys their atoms, although something of their material value is retained in the new molecule of water. For one kind of animal to live it must eat another kind of animal, or eat plants; and for plants to live they must use up the chemicals of the soil, water, and the energy of sunlight, or like fungi live on corrupted plants; etc.

Moreover, in a material world, things happen not only by deterministic natural law but by *chance*. Although each body by nature operates in a uniform way, its action can be frustrated by the natural action of another body, and this interaction can be a truly chance event, because it is not regulated by a higher cause. The fact that God in his infinite power can and has chosen ultimately to glorify the universe as a "new creation" in which all destructive events, even those that are matters of chance, will cease (and thus eliminate subhuman animal life and death) does not mean that it was evil of God to create the present universe where physical evils do so obviously exist. The creator did so, often for reasons that may not be evident to us (one of which is that a developmental universe would not otherwise be possible!), so that we humans at least might experience the present world in its variety, including chance events, and share with him in its completion and governance. Thus, for us, it is a greater good that human history has been played out in a world of good *and* evil that has given those who are willing to play a role in the drama the opportunity to do so for the sake of its ultimate marvelous outcome. Thus we have the opportunity to experience that drama, and from it come to wisdom perfected in faith here and now and in the beatific vision hereafter.

But what "here and now" of the pain of animals, which, as mentioned before, Peter Singer claims proves that they have rights?[11] Their pain warns them to remove their bodies from

---

[11] Singer is a Professor at Princeton; see Chapter 3 above, text p. 107 at note 85, and Singer 1975: *Animal Liberation: A New Ethics for our Treatment of Animals.* As a utilitarian in ethics, Singer argues that although humans suffer

injury, and generally succeeds. Furthermore, animal pain is very different than human pain, as psychologists have shown, because it is free from the *self-conscious* fear (especially of death) that makes human pain so great.

On the other hand, there is a sense in which *moral* evil is the only true evil, for it alone arises from freedom, whereas physical evil is an unavoidable side-effect of physical interactions among finite beings. For this reason, there had to be a human animal in order for there to be actual *sin* (i.e., moral as distinct from merely physical evil), and there had to be a "first" or "original" misuse of freedom by that human animal. It is the misuse by an intelligent creature of its share in God's freedom, when the creature's action opposes God's always good purposes *precisely as those purposes are at work in the human situation species-specifically distinguished by the presence of intellect and will.* Here is worth citing St. Thomas's profound insight that, as a consquence of the infinite goodness of the Divine nature, not even the moral law can be a matter of an arbitrary stipulation, for, contrary to the voluntarists and nominalists and to Muslim teaching that God can make anything good by decree of his will, moral good is precisely the good of human nature as created by God to come to positive expression in free human action:[12]

---

more pain than animals and thus have greater rights, still, since animals have various grades of pain, they have various grades of rights.

[12] Aquinas 1259/1265: *Summa contra gentiles* III, chap. 122: "*non videtur esse responsio sufficiens si quis dicat quod facit iniuriam Deo. non enim deus a nobis offenditur nisi ex eo quod contra nostrum bonum agimus.*" See John Deely 2010b: the *epigrams* to Chapter 8, "Albert (†1280) and Aquinas (†1274): Focusing The Challenge of Reason", p. 171, together with "*Addendum to Chapter 8:* Projecting into Postmodernity Aquinas on Faith and Reason".

Vincent Genovesi 1966: *In Pursuit of Love*, p. 109, suggests that "our whole perspective on God and sin would change" if the human good were commonly understood in the perspective that Aquinas presents, inasmuch as "we would understand that nothing is or becomes evil because it is forbidden by God's law", but rather the converse. Just as not even God can restore lost virginity or make a square circle, so not even God can make something "sinful" which is in line with the human good, and conversely.

If anyone says that something is a sin because it offends God, the thinking falls short. For God has so created human beings that it is impossible for us to offend God except by acting contrary to the human good.

Or again Aquinas:[13]

Since the good of human beings stems from reason as its root, this good will be the more perfect to the extent that it can be derived from consideration of the many things appropriate to human flourishing. Whence no one doubts that it pertains to the perfection of moral good that our outward actions be directed through the rule of reason ... in accordance with what is said in Psalm 83:3: 'My heart and my flesh have rejoiced in the living God': where by 'heart' we are to understand the intellectual appetite, and by 'flesh' the sensitive appetite.

Salvation is an individual matter in all circumstances, but the growth of reason is a collective enterprise.[14]

---

[13] Aquinas 1266/1273: *Summa theologiae* I-II, q. 24, art. 3: "*cum enim bonum hominis consistat in ratione sicut in radice, tanto istud bonum erit perfectius, quanto ad plura quae homini conveniunt, derivari potest. unde nullus dubitat quin ad perfectionem moralis boni pertineat quod actus exteriorum membrorum per rationis regulam dirigantur. ... secundum illud quod in Psalmo LXXXIII, dicitur, 'cor meum et caro mea exultaverunt in Deum vivum', ut cor accipiamus pro appetitu intellectivo, carnem autem pro appetitu sensitivo.*"

[14] The growth of reason in human civilization, at least in the Western regions where Christianity underwent the Protestant Revolt against the extreme abuses of clerical power in the Latin Church (particularly those centered on using the notion of indulgence to reap financial harvests), took its greatest symbolic blow historically in the Trial of Galileo, as we say in Chap. 2 above, *passim*, especially in our note on the main point of the whole affair (p. 41 note 10). The stunning effect of that trial led Descartes, and after him what was to become the early modern mainstream development of philosophy, completely to break with Scholasticism and to seek a new way to knowledge (see "The Crash and Burn of Scholasticism", Chap. 13 in Deely 2010b: *Medieval Philosophy Redefined*, pp. 381–384). But what was really at stake in the modern revolt? It was simply the realization that the study of nature could no longer base itself on books, but required experimentation, mathematization of results, and the use of instruments to make us aware of dimensions of

Salvation was a concern even when, as in St Paul's mind, the end of the world and Christ's "Second Coming" was less than a lifetime away. But we have learned quite certainly that early apostolic belief was a misreading of the signs, while "consideration of the many things appropriate to human flourishing" that St. Thomas calls upon us to ponder when speaking of sin has grown to an extent hardly imaginable.[15] It is as if the Incarnation had the purpose, beyond our salvation as individuals at the moment of death, of raising our species from its animal origins to its full spiritual dimensions of goodness — as seems suggested by Paul in *Romans* 8.22: "the whole creation has been groaning as in the pains of childbirth right up to the present time" — more than to restore a "lost innocence" in the mythical Eden of fundamentalism.[16]

------

physical reality quite hidden to inquirers relying wholly and solely on what can be known by the unaided senses. This transition to ideoscopy should have been a smooth and natural one, according to St Thomas's idea — no doubt evidencing the influence on his thought of St Albert the Great, a man who was interested precisely in science in the modern sense before that science existed — that the "natural philosophy" of Greek and Latin times required experimentation to reach its full maturation (e.g., Aquinas, c.1266: *Summa theologiae* 1.84.8c). Precisely for this reason Ashley 2006: *The Way Toward Wisdom*, p. 60, argues that "The time is ripe for a postmodern reconsideration of First Philosophy itself, now that the events of intellectual culture have brought us to the point where 'metascience' [rather than 'metaphysics'] can be seen to be the logically proper postmodern name for the recovery of this ancient enterprise in the spirit originally proper to it." See, in *The Way toward Wisdom*, the whole of Chap. 5, "The Existence and Essence of Metascience", pp. 132–169; also the review of this work by John Deely 2009c: "In The Twilight of Neothomism, a Call for a New Beginning. A return in philosophy to the idea of progress by deepening insight rather than by substitution".

[15] See the further discussion of human responsibility for the earth below at p. 144f., text and note 53.

[16] Thus, e.g., Glatz 2008 reports on an article by Fiorenzo Facchini, of the Department of Experimental Evolutionary Biology, Anthropology Unit, at the University of Bologna, Italy, which proposes that "rather than picturing [evolution] as humans descending from apes, ... humans ascended or rose up from the animal kingdom to a higher level" with the infusion of a soul able to cognize intellectually. Cf. "spiritualized animal" as mentioned above on pp. 80n29, 140–41, 158, 163.

Our point is that *on either interpretation, the demands of spiritual life remain the same*: "to observe the proper order in loving other things by preferring spiritual to bodily goods", as St Thomas put it.[17] The initial opposition of action to God's will by the first human couple is what the author of *Genesis* ascribes symbolically to a serpent (taken further to be a manipulation by a fallen angel) luring Eve and Adam in their choice to "be as gods, knowing good and evil". "To know good and evil" means to use one's power of freedom to do not only good but evil: deadly poison under the appearance of a kiss, like Judas' kiss of betrayal. God, like a good parent, can *permit* his children not only to do good but also to do evil, to sin, so that they will learn the consequences of sin and take responsibility for their freedom. If God had not freely done this he would have shown less respect for the genuine independence of those among his creatures which he has gifted with intelligence.

Thus, while God can cause *physical* evil, in that to create a material universe of interacting finite beings entails interrelations wherein the good for one is evil for another, he never causes *moral* evil, and both are permitted for a greater good of the whole. God places a material universe in existence, and *physical* evil results along with physical good in the interactions of the created finite beings, but with the overall result that the greater good of a varied and hierarchical universe came about. Moral evil he *permit*s to respect the freedom of his intelligent creatures. Thus he never permits evil, neither moral nor even physical, except as the occasion to some greater good that he intends, and ensures will be brought about. Of course, it is because we often do not recognize and cannot predict this good outcome that the great evil in the world shocks us, as it did Darwin; but both our reason and faith tells us, first, from the goodness of his creation that God exists,

---

[17] Aquinas c.1264: "On Reasons for Religious Belief", cap. 5: "... ut etiam in ceteris amandis debitus ordo servetur, ut scilicet spiritualia corporalibus praeferamus."

and second that, as Goodness Itself, he can and will bring about an overall greater good.

The Second Objection to Human Evolution

A second major objection to a theory of evolution is that it lowers human dignity. This is the case, however, only if one fails to understand that biological evolution can apply directly only to the human body. While the whole of the physical universe, body and soul in the human case, God creates from nothing, the intellectual soul God not only creates from nothing but *infuses* in the body — immediately creates — as the completion of his creation through evolution and the biological interaction sexually of the human parents.[18] As we have already shown, there are sound philosophical arguments concluding to the spirituality of the human soul as a substantial form that is not wholly reducible to the potency of matter. Speaking in 1950,[19] Pius XII made clear that this

---

[18] Recall Poinsot's demonstration above (Chap. 2, page 55, note 27) as to why the creation of the human soul yet belongs to the order of nature, despite its spiritual nature exceeding the pure potency of matter.

[19] Pope Pius XII 1950: *Humani Generis* par. 36: "the Teaching Authority of the Church does not forbid that, in conformity with the present state of human sciences and sacred theology, research and discussions, on the part of men experienced in both fields, take place with regard to the doctrine of evolution, in as far as it inquires into the origin of the human body as coming from pre-existent and living matter — for the Catholic faith obliges us to hold that souls are immediately created by God. However, this must be done in such a way that the reasons for both opinions, that is, those favorable and those unfavorable to evolution, be weighed and judged with the necessary seriousness, moderation and measure, and provided that all are prepared to submit to the judgment of the Church, to whom Christ has given the mission of interpreting authentically the Sacred Scriptures and of defending the dogmas of faith. Some however, rashly transgress this liberty of discussion, when they act as if the origin of the human body from pre-existing and living matter were already completely certain and proved by the facts which have been discovered up to now and by reasoning on those facts, and as if there were nothing in the sources of divine revelation which demands the greatest moderation and caution in this question."

was also the one qualification to current theories of evolution on which Church teaching insists.[20] Pope John Paul II,
speaking in 1996, re-emphasized Pius XII's point,[21] while

---

[20] Thus *Humani Vitae* in 1950 may be said to mark a theological turning
point. Garrigan 1967: "Theological Aspect [of Human Evolution]", p. 684,
well notes that, "for most of Christian history … the traditional idea that
God had created things as they are, fixed in species, did not have a serious
rival. Evolution (transformism) came into prominence chiefly through the
work of Charles Darwin" only in *1859*, and the main figures of the time who
took up the evolutionary view (such as Thomas Huxley in England and Ernst
Haeckel in Germany) "were militant materialists, atheists, or agnostics".
Churchmen accordingly tended to respond to the new idea — an evolutionary universe, including the human species — as to an attack upon religion in
general and Christianity in particular, with the unfortunate result of creating
an atmosphere of acrimonius rather than dialogical controversy.

It did not seem to occur to the churchmen of the time that, as Garrigan
notes (ibid.), "statements of the Fathers and theologians before Darwin are
not strictly *ad rem*, since the issue had not even been raised" in the earlier
centuries. "Vatican Council I in *1870* was content to repeat the common-
sense advice: the same God gives revelation and reason; one truth cannot
contradict the other (Denzinger 3017)", Garrigan notes. **In *1893*** Pope Leo
XIII applied the principles earlier enunciated by Augustine "to cases of apparent conflict between science and the Bible: Whatever they [scientists] can
really demonstrate to be true of physical nature, let us show to be capable of
reconciliation with our Scriptures (Denzinger 3287)".

Still, "**in *1909*** the Biblical Commission refused to call into question the
literal and historical meaning of Genesis". So, even though, as Nogar reports (1966: 115–116), Pope Pius XI — "to his great credit, and to our edi-
fication" — when "asked [in *1938*] by his theological advisors to use his authority to put an end, once and for all, to the 'insidious' doctrines of physical
and cultural evolution … he is reported to have replied: 'One Galileo case in
the Church is enough'," it was really not "until Pope Pius XII [in *1950*] rel-
egated the statements of the Holy Office about physical origins to the status
of historical interest, the organic origin of man was not an open question
in theological circles." A pity these early-20th-century-late-modern clerics
had not familiarized themselves with the anticipatory postmodern view on
the question adumbrated by Poinsot in 1643 (as cited above in Chap. 2,
page 56, note 27) in line with the understanding of the cenoscopic physics
of Aquinas and Aristotle! Only in *1996*, with Pope John Paul II's "Message
on Evolution", would modern theology fully catch up with the theology of
John Poinsot on the connection between the spirituality of the human soul
and the material universe within which it provides a principle of speciation.

[21] Pope John Paul II 1996: "Message on Evolution", pars. 5 & 6: "Pius

further[22] remarking that[23]

> Today, nearly a half-century after the appearance of that encyclical, some new findings lead us toward the recognition of evolution as more than an hypothesis. In fact it is remarkable that this theory has had progressively greater influence on the spirit of researchers, following a series of discoveries in different scholarly disciplines. The convergence in the results of these independent studies — which was neither planned nor sought — constitutes in itself a significant argument in favor of the theory.

The Third Objection to Human Evolution

The third objection to human evolution is that it reduces human origins to chance. It is true that Neo-Darwinism explains

---

XII underlined the essential point: if the origin of the human body comes through living matter which existed previously, the spiritual soul is created directly by God ("*animas enim a Deo immediate creari catholica fides non retimere iubet*" — *Humani Generis* par. 36).

"With man, we find ourselves facing a different ontological order — an ontological leap, we could say. But in posing such a great ontological discontinuity, are we not breaking up the physical continuity which seems to be the main line of research about evolution in the fields of physics and chemistry? An appreciation for the different methods used in different fields of scholarship allows us to bring together two points of view which at first might seem irreconcilable. The sciences of observation describe and measure, with ever greater precision, the many manifestations of life, and write them down along the time-line. The moment of passage into the spiritual realm is not something that can be observed in this way — although we can nevertheless discern, through experimental research, a series of very valuable signs of what is specifically human life. But the experience of metaphysical knowledge, of self-consciousness and self-awareness, of moral conscience, of liberty, or of aesthetic and religious experience — these must be analyzed through philosophical reflection, while theology seeks to clarify the ultimate meaning of the Creator's designs."

[22] So we reach another turning point, one anticipated in Garrigan's remark (1967: 685) that "the view is beginning to emerge, inverting the common opinion of the last century, that revelation says less about evolution than evolution says about the theology of creation."

[23] Pope John Paul II 1996: "Message on Evolution", par. 4.

evolution by "natural selection" of chance mutations in the genome of a species of organisms, and the origin of life to some
kind of accidental occurrence in inanimate matter — e.g.,
the Miller-Urey theory that life was produced by lightning
striking a chemical "soup" in shallow waters.[24] One of the most
eloquent defenders of Neo-Darwinism, Stephen Jay Gould
(1941–2002), was insistent that the cause of evolution was a
sequence of chance events that might as well, or even more
probably, have ended not in intelligence but in some species of
bacteria or insect.[25] Of course it was suggested, as long ago as St.
Augustine, that there might have been an innate determinism
in matter to produce higher forms.[26] The popular understanding of evolution often takes this form, and it was supported
by the "creative evolution" of the philosopher Henri Bergson
(1859–1941),[27] as also by the writings of the paleontologist/
priest Pierre Teilhard de Chardin (1881–1955) with his idea of
God as the "Omega Point" of the evolutionary cosmos.[28]

Current science, however, provides no evidence for any
such general or overall "law" of evolution. This is the main
defect of Neo-Darwinism: it is a scientific theory that explains
observed facts (mainly the fossil record of successive forms of
life) ultimately by natural selection based on chance mutations,
while leaving out the finality of the individual organisms as

---

[24] See "Abiogenesis" at <http://en.wikipedia.org/wiki/Abiogenesis>; and
White 2007: "Abiogenesis — Origins of Life Research" at <http://darwiniana
.org/abiogenesis.htm>.

[25] See Gould 1997: "The Evolution of Life on Earth" at <http://eddiet-
ing.com/eng/originoflife/gould.html> (under <http://eddieting.com/eng/
index.html>). Gould's claim, it should be noted, rests on thinking in exclusively ideoscopic terms; for a cenoscopic analysis of the same material (Deely
1969) demonstrates the opposite conclusion.

[26] See the *Wikipedia* entry "Intelligent Design" <http://en.wikipedia.org/
wiki/Intelligent_design>; and Jonathan Wells 1998 shows that St. Augustine
did not believe in the evolution of species, but rather that all species were
created in seed (*rationes seminales*) to emerge in sequence over time.

[27] Bergson 1907: *L'Évolution créatrice*.

[28] See esp. Teilhard 1959: *The Phenomenon of Man*.

giving directions to chance factors which precisely influence the over-all direction that "natural selection" effects.[29] Like cosmological theories that explain the Big Bang by saying "it just happened", this contradicts the first principle of science cenoscopic and ideoscopic alike: namely, that nothing changes without a cause. Stuart J. Kauffman recognizes this when he hypothesizes that living things have an innate power of "self-organization" through catalysis, but honestly admits that this is as yet only a hypothesis.[30]

Furthermore, the evidence for the fact of evolution of species, today "more than a hypothesis", as Pope John Paul pointed out,[31] is based on two observations, the first of which alone is decisive. For the two observations belong to two different orders or levels of understanding: the first concerns what Aquinas called a *demonstratio quia*, or proof *that* something occurred; the second concerns what Aquinas called a *demonstratio propter quid*, or a proof *of the reasons for the fact*, the identification of the actual causes.

The first observation on which evolution is based is the fossil record, although this record has many gaps, and especially the relatively sudden appearance of all the main phyla of animals in the Cambrian period, called by Stephen Jay Gould "punctual equilibrium". It can be said that all the fossil record

---

[29] See Waddington 1960: "Evolutionary Adaptation", and 1961: "The Biological Evolutionary System"; also "The Framework of Evolutionary Science" in Deely 1969: 102–130, esp. 105–111.

[30] Stuart J. Kauffman 1995: *At Home in the Universe: The Search for the Laws of Self-Organization and Complexity*. Kauffman, without realizing it, seems to be pointing here to the very problem biology has wrestled with since Descartes' removal from natural analysis of any notion of "intrinsic final causality": see "A lair for later nonsense: from Teleology to Teleonomy", in Deely 2001: 65–66. This goes back to the shortcoming in common evolutionary theory pointed out by Waddington 1960 and 1961, and wrestled with also by Colin S. Pittendrigh 1958 in his coining of the expression "teleonomy" to plug the Cartesian hole, at least from within biology.

[31] Pope John Paul II 1996: "Message on Evolution", par. 4, cited on p. 127, text at note 23. See the following pars. cited in note 21, pp. 126–127.

certainly proves is that there has existed a series of life forms, most of which have now perished, sometimes in massive extinctions due to some kind of natural disasters, and that the now living life-forms have been preceded not merely by *other* forms but by typically and *specifically different* forms no longer existent. We say that this much is decisive for accepting evolution, because proof of this much leaves us with only two options: *either* we can proceed on the assumption that the present state of the world is causally connected with past states, and so the relation between the two — which is what is meant by "evolution" in this case — can be investigated; *or* we can assume that there is no natural connection between presently existing life-forms and those which existed in the past, that God simply and arbitrarily created and extinguished species leaving no traceable reason — on which assumption there is no point whatever to science and philosophy, and no rational component to be reckoned with in theology. It is the first assumption — that the present state of the world emerges causally from past interactions of life-forms among themselves and with their environment — that is the basis presupposed for any science, cenoscopic or ideoscopic.[32]

The second kind of evidence for evolution is that we observe species to vary within themselves as a result of environmental variations; and there are some cases, such as the several species of finches that Darwin noted on the Galapagos Islands, where adaptive variations seem to have become specific differences in relatively short times. The weakness of this argument is that the main definition of "species" used in biological science today is the infertility of cross-breeding between two varieties or types. While this is certainly evidence of speciation, it is not a conclusive criterion, since what makes one animal species essentially different from another should be found in its highest

---

[32] See "The Impact of Evolution on Scientific Method" in Deely and Nogar 1973: *The Problem of Evolution*, pp. 3–82, esp. 21–22.

powers to act, which would mean in its psychology;[33] and to date this sort of animal taxonomy remains very incomplete.

Current science has, as we have seen, tended to dogmatically close off the possibility that the universe includes not only extraterrestrial embodied intelligences but also pure spirits. If this possibility is considered, then an evolutionary theory becomes possible based on the analogy of evolution (used in *Genesis*, but not dependent on revelation) between a human craftsman and his work. This is not the late-modern "Design Argument" as it is usually presented, since that argument explains evolution simply as the act of God, while in this hypothesis, while God is the First Cause of existence, he acts through his creatures as a craftsman using tools. We know today that a chemist can produce more complex chemicals than those that occur in nature in the absence of life. Therefore there seems no reason that it is impossible that eventually we may be able to produce living things in the laboratory using natural forces. Many have tried unsuccessfully to do so since Miller and Urey attempted it in 1952,[34] but that does not prove it is impossible. Scientists are willing to admit that perhaps extraterrestrials more advanced in intelligence have already succeeded. What, then, is unscientific about supposing cenoscopically, as Ashley has argued elsewhere,[35] that God may have put this task in the hands of pure spirits, the angels of biblical fame? This would not make them creators, but servants of the Creator, as our technologists are servants in "cultivating our garden earth". One consequence of such a hypothesis would also be that pure spirits who have become

---

[33] See the discussion of the contrast between and consequences of the species notion in the Thomistic philosophical tradition vis-à-vis the species notion as entertained by modern science in Section I, "The State of the Question", Deely 1969: 76–90, esp. 79–90.

[34] See the *Wikipedia* entry "Miller-Urey Experiment" at <http://en.wikipedia.org/wiki/Miller%E2%80%93Urey_experiment>.

[35] Benedict Ashley 2006a: "The Existence of Created Pure Spirits", Chapter 4 of *The Ashley Reader*, pp. 47–59.

132 How Science Enriches Theology 🍃 Benedict Ashley and John Deely

demons by misuse of their free will have also been attempting to distort evolution (although they are opposed by the good angels, who will eventually win out). This would be an additional factor in answering the problem of evil, since it would mean that not only human life but the whole universe would bear the mark of a dualism, such as was favored by Zoroastrians and Gnostics, although this dualism would only be phenomenal and not ultimate or contrary to the fact that God made everything very good.

What then of the Design Argument, being so vigorously debated at the present? It is not correct to claim that this Argument is a religious intrusion on rational science, inasmuch as the argument has been formulated by Michael Behe,[36] as one of its chief proponents, purely on the basis of microbiology. Behe's first claim is that living organisms are "irreducibly complex" because mutations in their genome cause what must be rated as genetic defects, and thus render them less adapted to survival. His second claim is that in fact no microbiologist has been able to propose a detailed hypothesis as to how life originated or one species has evolved into another, especially a more complex one, by a series of genetic changes that would produce fertile offspring.[37] Third, he notes that the Anthropic Cosmological Principle, even in its weak form, shows that the existence of life on our planet (and if, as many claim, on other planets) is much too improbable to be explained by chance.

---

[36] See Behe's webpage, <www.arn.org/authors/behe.html>, with answers to critics and bibliography (part of the "Access Research Network" <http://www.arn.org/>).

[37] It is hard to see this argument as anything other than an inexplicit version of the generally discredited "God of the gaps" approach: whatever we cannot at time X explain "must" be the direct action (or intervention) of God. See the entries "God of the gaps" in *Wikipedia* <http://en.wikipedia.org/wiki/God_of_the_gaps> and also in *Theopedia, an encyclopedia of Biblical Christianity* <http://www.theopedia.com/God_of_the_Gaps>. Compare the more careful approach formulated in "God, the Designer of Evolution", Nogar 1963: *The Wisdom of Evolution*, 387–395.

Although attempts are being made to answer these arguments, they remain very strong.[38]

Nevertheless, in our opinion, theologians should treat the late modern versions of the Design Argument with caution, because the biological and the cosmological arguments on both sides of the debate depend on an elaborate mix of fragmentary data and shaky hypotheses, while the basic proofs of God's existence (especially the "first and more manifest way" from our experience of motion) as formulated by Aquinas rest on much simpler, general, and more certain sense data. Although Aquinas' basic arguments are apparently supported by the Design Argument, they are in fact quite independent of it. The First Way through motion discussed in Chapter 1 remains, as Aquinas says, the most evident, since it rest on nothing more than the fact that our universe exists in the process of change, a fact that is presupposed by all of science and everywhere evident to sense. The Design Argument, as it is being formulated today, by contrast, combines the Fourth Way through formal causality and the Fifth through final causality.[39] Hence it *presupposes* (for whatever

---

[38] But note that the "strength" of Behe's argument depends mainly on the lingering presence of the Enlightenment view that science concerns only what can in principle be seen and touched, and hence is exclusively ideoscopic; for when the cenoscopic dimension underlying and providing the framework of intelligibility presupposed for the initial development of ideoscopic science is expressly brought into the picture, Behe's line of argument is seen to beg the question of the ideoscopy/cenoscopy distinction, in the sense of what question properly belongs to which type of science, or in what admixture?

[39] The Fifth Way, the "proof from governance", it should be noted, is the traditional argument to which the current expression "Design Argument" most readily transfers; and yet this fifth proof is the only one that, as expressed by Aquinas in his *Summa theologiae* I, q. 2, art. 3, stands on the shaky ground of what has proved to be the myth of the unchanging heavens: see "The deficiency of the 'Fifth Way', and the matter of alternative further 'Ways'," in Deely 2010b: 190–191; see also the extended treatment in Christopher Martin 1997: "The Fifth Way", Chap. 13 of *Thomas Aquinas. God and Explanations*, pp. 179–206.

Feser 2009, though blandly dismissive (pp. 119–120) of the serious defect in the 5th Way as stated in Aquinas' 1266 *Summa* text, does point out

validity it can achieve in its own right) the first three of the Five Ways based on efficient causality, since final causality is the pre-determination or directedness of efficient causality, whether it is natural and deterministic or free. Thus the traditional proofs for the existence of God as presented by Aquinas can be understood in a more fundamental way than can be the Design Argument as it is now being proposed. Yet as supporting considerations, perhaps, certain formulations of "Design Argument" may none-theless help to enrich our view of the Creator, and especially of the Holy Spirit as the guide of the universe's history.

We can conclude therefore that theologians can accept the fact of evolution, and even the hypothetical account of its causes given by Neo-Darwinism as far as it goes, but should at present remain skeptical as to how complete an explanation of this fact Neo-Darwinism has so far supplied in ideoscopic terms. What philosophic analysis in cenoscopic terms adds to the post-Darwinian picture of evolution as understood today, however, more than justifies the continued insistence by theo-logians that (a) the origin and evolution of life, like all physi-cal change, is impossible without a First Uncaused Cause; and that (b) the substantial forms (or "souls") of human animals, even though involved in evolution, require an aspect of effi-cient causality directly from the Creator in that they are not wholly drawn from matter but are directly created (as would also be true necessarily for pure spirits). Thus theologians should support continued research for a better understanding of the evolutionary process, even while maintaining caution about claims, such as those of Teilhardians, once so popular among theologians, that there is a deterministic law of evolu-tion evident at the level of ideoscopic science.[40]

---

correctly (p. 110ff.) how the "5th Way" differs from what early 21st century authors have called "Design Argument". (See Himma 2009 <http://www.iep.utm.edu/design/> for a general discussion of the point.)

[40] See "The Strange World of Fr. Teilhard", in Nogar 1966: *The Lord of the Absurd*.

Is the Holy Spirit Freedom?

In previous chapters it has already been shown that modern science can enrich our understanding of how the intelligence and free will that are properties of the human spiritual soul necessarily climax the hierarchy of the material and spiritual beings that constitute our created universe as a unified system of interacting finite beings. Freedom transcends both the determinism of natural laws and the chance events that result from the accidental interactions of material substances, because freedom is based on the spiritual intelligence in its practical exercise. Practical intelligence reasons concerning the means in relation to the innate goals of things, and above all (ideally) under the realization that God is the ultimate goal or final cause of all reality. Hence free will either freely directs the proper choice of means to those goals, or freely destroys itself by leading the creature to sin by acting falsely, as if it could make itself this ultimate goal. As the serpent said to the woman to persuade her to taste the forbidden fruit (Gn 3:5), "Your eyes will be opened and you will be like gods who know what is good and what is bad". But the freedom God gave us to seek our true goal is guided by the Third Person of the Holy Trinity in fulfillment of the will of the First Person, God the Father, and after the pattern of the Second Person, the Son and Word Incarnate.

It was clear in the history of the Church's guidance of its members, from Pentecost on, that this guidance is not merely legal and external, as in the Old Testament Law given to Moses, or as is claimed in the Qur'an.[41] Instead, it is the internal law of prudent intelligence and free and deliberate choice motivated by love of God and neighbor, as Jesus taught in the Ser-

---

[41] For the views of Thomas Aquinas on Islam, besides the discussion of sin as offensive to God only because it is action contrary to the human good, cited above at p. 121, see Aquinas i.1259/65: *Summa contra gentiles* Book 1, chap. 6. n. 7; Deely 2010c: "Projecting into Postmodernity Aquinas on Faith and Reason", 279–301; and Deely 2011: "Taking Faith Seriously".

mon on the Mount. The fact that the Jews and, many times in the course of history, Christians in the preaching of the Gospel and its catechesis lapsed back into a voluntaristic, legalistic emphasis only illustrates the sinful condition that permeates human interactions. Modern psychology has shown that children move from a purely animalistic and consequentialist understanding of morality ("If I grab that delicious cookie, my mother may slap my hand"), through a conformance to peer approval ("That is what other kids are doing, so I should too"), to a prudential understanding of the relative importance of human needs ("I have to sacrifice lesser to greater goods"). Freud's notion of the second step of this moral development as the subconscious Super-Ego is one of his better clinical insights. We argued in Chapter 3 that to make a free decision we need an appropriate image in the evaluative sense (*vis cogitativa*), and if the Super-Ego is distorted (a principal consequence of the human condition as sinful) this can make prudent action very difficult. Psychotherapy, therefore, should seek this healing of subconscious wounds to permit truly free and realistic decisions.[42]

The will of intelligent creatures is free because their intelligence shows them that any good that they know is finite, and hence that under the negative aspect of it as a limit it does not perfectly attract or satisfy. For us even God, the Infinite Good, is only imperfectly known, and hence seems to have limits. With the Trinity, however, the Father and Son love each other by the Holy Spirit as infinitely good and perfectly fulfilling, but the One God is free with respect to creation, since He needs nothing outside the Divine Goodness. Thus it is an error of some philosophers and theologians to attempt to *deduce* from the Neo-Platonic principle "the good is diffusive of itself", which is indeed true, any conclusion about what God, because he is good, "must" do outside the Divinity itself.

---

[42] For more on this, see Ashley 2011: *Healing for Freedom: A Christian Perspective on Personhood and Psychotherapy.*

Norman Kretzmann, in his fine commentary on the *Summa Contra Gentiles*,[43] has (we think mistakenly) argued that God's freedom with regard to creation concerns only *what* he creates, but nevertheless because he is the Good he *must* create something. Some Franciscan theologians have argued from the same principle that the Son *had to* become Incarnate in order to perfect the universe.[44] St. Thomas Aquinas, however, while admitting the fittingness (*convenientia*) of the Creation and Incarnation, refuses to attempt to *deduce* these truths. God needs nothing outside the Godhead, and Goodness itself need not will anything outside the Divinity. The only way that God's will is "determined" is that he can neither directly cause, nor even permit, any evil in creation except as instrumental in the achievement of a greater good of the whole creation.

This theological analysis of the analogy between the freedom of intelligent creatures and God's absolute freedom is enriched by the scientific psychology of human freedom and its relation to the determinations of the human body, especially of its brain, nervous and hormonal systems. It is further supported theologically by the arguments already given concerning the putative role of pure spirits in the governance of the world and disruption of that governance by angelic sin. Most importantly, it is enriched by elaboration of the wonderful system of the universe, in contrast to its finite limits and possible ultimate decline to be described later. Thus science has greatly weakened both the early modern steady-state hypothesis and the ancient notion of an eternal universe moving infallibly in cycles.

In considering the work of the Holy Spirit in guiding the universe to its goal, however, another facet of the problem of evil arises. A special feature of the symbolic narrative of the

---

[43] Norman Kretzmann 1999: *The Metaphysics of Creation: Aquinas's Natural Theology in Summa Contra Gentiles II.*

[44] For example, Seamus Mulholland, OFM 2007: "Incarnation in Franciscan Spirituality".

creation account of *Genesis* 1–3 is the intimate relation that the first human persons have with God. Although it does not say that they see God, nevertheless He speaks to them and they reply as if they were present to each other. After recounting the "original sin", the story continues (Gn 3:8–13):

> The man and his wife, when they heard the sound of the Lord God moving about in the garden at the breezy time of the day, hid themselves from the Lord God among the trees of the garden. The LORD God then called to the man and asked him, "Where are you?" He answered, "I heard you in the garden; but I was afraid, because I was naked, so I hid myself". Then he asked, "Who told you that you were naked? You have eaten, then, from the tree of which I had forbidden you to eat!" The man replied, "The woman whom you put here with me — she gave me fruit from the tree, so I ate it". The Lord God then asked the woman, "Why did you do such a thing?" The woman answered, "The serpent tricked me into it, so I ate it".

After God expels Adam and his wife from the Garden their intimacy with God ceases. The Fathers of the Church interpreted this to mean that we still possess the nature God gave us (although it is somehow weakened by the effects of sin, alike of original sin and our own sins), but we have lost that original state of intimacy with God. The Protestant Reformers, the Catholic Michael Baius and the Jansenists exaggerated this "weakening" of human nature, so that it amounted to its essential corruption[45] — a view contrary to reason, in that a completely corrupted yet existent "nature" is a contradiction in terms. Yet the Reformers were able to find support for this antinomic view in statements of some among the Church Fathers — most especially St. Augustine — statements, it must be said, that remained theologically vague. A clear distinction of

---

[45] Luther, for example, compared "human nature" redeemed by grace to a pile of dung covered over with snow!

grace from nature was, however, finally achieved by Aquinas, and has received acceptance in Church teaching.

Yet the problem was again raised with good intentions by a patristic theologian of the Vatican II period, Henri de Lubac, S.J. (1896–1991), who understood the received view of Aquinas to be an unsatisfactory "two-story" theory in which the human person has a "supernatural end" imposed by grace on top of its "natural end".[46] De Lubac attempted to replace this by asserting that we have one end only, our supernatural end of union with God by grace. Then, in order to save the gratuity of grace as a pure gift of God, de Lubac still maintained that we cannot attain this graced union by God by any effort of our own. Thomists were quick to point out that for Aquinas this would be a flat contradiction, since a "nature" is defined by its formal cause, and in the substantial order a formal cause without a correlative final cause is impossible.[47] To say that God created a human person is to say that He gave that person a complete nature able to attain its end by that person's own independent powers (although, of course, not without God as First Uncaused Cause, i.e., source of all actual existence here and now).

Is the Holy Spirit Generous?

The Bible speaks of God as a "fountain of living waters" (Jer 2:13; 7:13); "For with you is the fountain of life, and in your light we see light" (Ps 36:10). This "flowing out" of God's inexhaustible *plenitude* in generosity is especially to be attributed to the Holy Spirit. Science has shown us an expanding universe in which vast galaxies of stars arise; and by the theory of evolution science has shown how subsequently thousands of species of living things have arisen in the waters, the deserts, and the

---

[46] See Henri de Lubac 1946: *Surnaturel*. This work itself was never translated, but an expanded 2-volume version appeared in English in 1967: *The Mystery of the Supernatural*, and 1969: *Augustinianism and Modern Theology*.

[47] See the two-part article by Antoninus Finili 1948, 1949: "Natural Desire"; also his 1952: "New Light on Natural Desire".

mountains of the earth. Human history is a narrative of this multiplication in the case of the intelligent human race and its remarkable ability to create technologies and cultures.

Christians believe, moreover, that the divine generosity extended to giving the first humans not only a wonderful spiritual nature, but to elevating that nature by *grace*, a term that in Latin means (1) what is pleasing, and (2) thanks for a gift (because it is pleasing). Thus grace is a generously given and excellent gift of the Holy Spirit. Aquinas, in discussing grace and nature, shows that God created humanity with a nature having as its ultimate end a *natural* union with God as Creator, but that this nature was also *intrinsically* elevated in its intimate being by grace to *supernatural union* with God as Trinitarian, as personal. What De Lubac found objectionable in this was, it would seem, that some later scholastics understood this elevation as *extrinsic* to human nature, although Catholic tradition has always held that the human person and all spiritual beings are *capax dei*, open to God, simply because they are spiritual in soul. That human persons are animals does not shut them off from God, precisely because they are *spiritualized animals.* This expression makes quite an important point, which needs to be explained as follows.

The Neoplatonic view under which Augustine mainly thought, and which permeated the intellectual Latin culture of Aquinas' youth and after, considered the human being as *an incarnate or embodied spirit*, sharing with the angels above an intellect whose existence did not depend upon the body or, for that matter, anything in the material order. So the contrary view that Aquinas expresses, when he addresses the question of whether the human soul pre-exists the human body,[48] amounts to a revolutionary shift, more important prospectively, perhaps, than it was at the time of a universe conceived on all hands as unchanging and centered on earth. To

---

[48] Aquinas 1266: *Summa theologiae Prima pars*, q 90, art. 4.

the assertion that "the human soul is more like the substantial form of angels than that of brute animals", Aquinas
responds to the contrary[49] that "if the human soul by itself
were a natural kind, it would indeed be more like the forms
of angels than like the forms of brute animals; but because it
is the form of a body, the human soul achieves the speciation
of a natural kind only as formal principle of an animal body,
and therefore within the genus of animal". In other words,
the human soul is not as much the embodiment of a spirit as
it is rather the formal principle *whereby the genus of animal
itself becomes spiritualized* through the human species.[50]

Even after sin, thus, the natural capacity of human nature
for a union with God in grace remains uncorrupted. Yet it is
weakened by sin, because wrong choices freely made create a
disharmony with either the natural or supernatural end of human existence. The amazing truth, known only by revelation,
is that God, in giving us this nature, also in his utter generosity willed to transform and elevate it to that intimate union
with God symbolized in *Genesis* 1–3. Thus human finality has
been raised not to simple rational contemplation of God in
his created effects, such as Aristotle considered, but to an intimate union with God through the grace of faith, hope, and
charity, directed ultimately to the beatific vision of God's in-
nermost being as Triune and Creator of all. If God had not
willed this transcendent grace, even apart from questions of
sin, it would still have been infinitely beyond our natural powers to attain the beatific vision. This does not mean, as De
Lubac thought, that grace is laid "on top of" nature, but that
our nature was created with an ordination to a transformed
and elevated state that penetrated our nature to its very depths.
To use an analogy, a mountain climber may reach a high peak,
only to discover that above it emerges a still loftier view, and

---

[49] Ibid. q. 90 art. 4, reply to objection 2.

[50] See the remarks of Facchini as reported by Glatz 2008 in note 16 of this
chapter, p. 123 above, with the "spiritualized animal" cross-references.

then find that surprisingly he has enough oxygen left to attain that further height.

What *Genesis* symbolizes and St. Paul repeatedly encourages us to understand is that what the Gospel promises is not only the healing of our human nature as enfeebled by sin, but also the recovery of that intimacy of grace that the first Adam possessed before sin and that in the second Adam, Jesus Christ, we can recover through faith, hope, and charity. As Pope Benedict XVI has summarized:[51]

> If, in the faith of the Church, an awareness of the dogma of original sin developed, it is because it is inseparably linked to another dogma, that of salvation and freedom in Christ. ... we must never treat the sin of Adam and of humanity separately from the salvific context — in other words, without understanding them within the horizon of justification in Christ.

Modern science, of course, knows nothing of grace as such, since that transcends human reason and belongs to faith alone. What science does show in ever greater contrasting details is the paradox of the marvel of the universe and human nature as we observe them, yet at the same time the presence, that so troubled Darwin, of so much physical and moral evil in the midst of the marvelous order and goodness of nature. Thus, our own experience deepens the problem of evil that the doctrine of grace alone can perfectly answer.

The analogy in the fine arts and in the technologies of "creative" human freedom (which presupposes something existing upon which to act) to God's truly creative freedom (as presupposing nothing) is especially instructive. The fine arts are practical in that, unlike theoretical science, they *produce* a material product — a painting, statue, music; and in literature with words that we at least imagine as audible — but their *goal*

---

[51] Pope Benedict XVI 2008: "The Apostle's teaching on the relation between Adam and Christ", 2nd par.

is not practical, but *contemplative*, as we explained in Chapter 3 in defending their mimetic character. Historically, mathematics and mathematicized science began with Pythagoras and his theory of music. An artist-inventor such as Leonardo da Vinci illustrates this close relation yet distinction between technology and art. Christian theology has rejected iconoclasm as a heresy,[52] and promoted both the sacraments and the fine arts in worship. The Holy Spirit breathes through these sensible symbols, healing and elevating the human soul. If, as urged in Chapter 3, artists would turn more to scientific discoveries, as a few have done, to understand the beauty of nature, instead of rejecting representation as mere copying, the fine arts could enjoy a "renaissance".

The technologies which have made such amazing advances through science can also enable us to enrich theology. The human authors of *Genesis* 1–3 rightly concluded that God must have created the human race in an environment that would support their nature and further its innate goals, and hence called it a "garden" that supplied all their material and esthetic needs. This agrees with the predominant scientific view which holds that the human race in its present varieties originated in East Africa less that 200,000 years ago, and did not spread far out of that area until about 50,000 years ago. This seems to indicate that the human family only grew slowly. This slow advance also fits the notion of original (and subsequent) sin that for long darkened human progress, and was further exaggerated when the race scattered through Asia and Europe and then into the Pacific and the Americas.

While this expansion furthered cultural variety, it also produced social isolation and technological regression. Finally, around 10,000 BC, agriculture and civilization arose in the Near East, but the development also produced its own social

---

[52] On the various factors that promoted this controversy, see the excellent article by Nick Trakakis 2004/5: "What Was the Iconoclast Controversy About?"

division and conflicts (symbolized by the Tower of Babel), with a decline of religion into a welter of mythology and idolatry.

Only with the so-called Axial Period, from about 700 BC to 300 BC, religious reformers — such as the writers of the Upanishads and Siddhartha Gautama the Buddha in India; Zoroaster in Persia; the Hebrew Prophets in Israel; Confucius and Lao Tzu in China; Plato and Aristotle in Greece — reacted against these cults and proposed highly systematic monist or monotheist "philosophies" leading toward the Gospel. The work of the Holy Spirit, more or less distorted by sin, is manifested in this global movement which modern science and historical research has made more evident. That work is reflected also in the growth of science and scientific technology in these cultures.

According to *Genesis* 1:28, God said to the first couple, "Be fertile and multiply; fill the earth and subdue it. Have dominion over the fish of the sea, the birds of the air, and all the living things that move on the earth". This has been taken to justify abusive dominion over the environment, but in *Genesis* 2:15 we read: "The Lord God then took the man and settled him in the Garden of Eden, to cultivate and care for it". The two terms "care" and "cultivate" indicate that we have a responsibility to conserve the environment, while at the same time making a proper use of it through technology.[53] Farming and harvesting are positive symbols in the Bible, and Jesus chose to work most of his life as "carpenter" or "builder", since the Greek in Mk 6:3, *tekton*, is the same root as our "technology", and has the wide sense of "practical reason".[54]

---

[53] See "Dominion or Stewardship?", in Ashley 2006a: *The Ashley Reader*, Chapter 16, pp. 271–290. Also John Deely 1969a and 2008a: "Evolution and Ethics" and "Evolution, Semiosis, and Ethics", reprinted in Cobley ed. 2009: *Realism for the 21st Century. A John Deely Reader*, "Section II. Ethics", pp. 54–90. See further Petrilli and Ponzio 2003: *Semioetica*; and the *Sequel* to Deely 2010a: "The Ethical Entailment of Semiotic Animal, or the Need to Develop a Semioethics", pp. 107–125.

[54] St. Thomas, in this following Aristotle (c.330 BC, *De Anima* Book 2),

The great abuses of current technology, such as the threat of atomic warfare, the wasting of energy resources, the heating of our atmosphere, the decrease of biodiversity, the pollution of water, and the aging of the human population, all have moral roots. At the same time, the many good uses of technology, such as the overcoming of famine, the increase of health and longevity, the facilitation of human communication, etc., are also rooted in good human striving to "care for and cultivate the garden" God has given us. Oddly, some scientists, such as Raymund Kurzweil, make such startling predictions of technological progress as the following:[55]

> Within a few decades, machine intelligence will surpass human intelligence, leading to The Singularity —

---

regarded speculative understanding — i.e., the ability to consider the nature of things over and above, apart from, their animal relation to us as desirable, dangerous, or safe to ignore — as the species-specifically distinctive cognitive power setting human animals apart from brute animals, which have only the awareness of sensory experience interpreted in just those terms (+, –, 0) of objects only in relation to themselves. See Deely 2002: *What Distinguishes Human Understanding?* Thus, while all animals know things objectively, only human animals can further come to know the physical surroundings in terms of the subjective constitution which things objectified have quite independently of their being objectified (e.g., all higher animals "know water when they see it"; but only human animals have or could come to know that H_2O is the formula that expresses "what makes water be water"). And this knowledge in turn provides a basis from which the human animals are able further and further (as the speculative understanding deepens) to extend their control over their surroundings. This is the root cause of the need to develop an environmental and not just a social ethics — that is to say, a "semioethics", wherein the human animal develops indeed a truly "global" responsibility (in the term established mainly by Sebeok, e.g., Sebeok 2001). See Aquinas 1266: *Summa theologiae* I, q. 79, art. 11, sed contra: "the speculative intellect becomes practical by extension; but it is not the case that one power is changed into another, for which reason speculative and practical understanding are the work of the one same cognitive power" which is species-specific to human animals as what sets them apart from all other animals on earth, gives them dominion — and, as it turns out, global responsibility.

[55] Ray Kurzweil 2001: "The Law of Accelerating Returns", opening paragraph. See further "Accelerating Change", *Wikipedia* entry <http://en.wikipedia.org/wiki/Accelerating_change>.

technological change so rapid and profound it represents a rupture in the fabric of human history. The implications include the merger of biological and nonbiological intelligence, immortal software-based humans, and ultra-high levels of intelligence that expand outward in the universe at the speed of light.

In *Theologies of the Body: Humanist and Christian*,[56] Ashley took up Karl Marx's futurist view that "someday man will create himself", a view with which Kurzweil is agreeing, and showed that it is more probable that our abuse of technology will end by destroying us than by granting us immortality. Similarly, Nick Bostrom, an Oxford philosopher, is one of the organizers of the "World Transhumanist Association" which hopes to make humanity immortal and super-intelligent.[57] Yet advance in practical, technological truth, when it is truly practical and life-promoting rather than abusive, like advance in the theoretical truth of pure science, is the work of the Holy Spirit both as to truth and as to motive; while its abuse adds to the heavy weight of accumulating original sin.

Because of human freedom, the invention of technologies differs from innate human needs as biologically and psychologically determined, in that there is no limit to good technologies we can devise. Yet, since technologies to be good ought to serve genuine human needs, and not disserve true human needs in unethical ways, we ought not to ignore the danger that an overly technical culture can gradually cut us off from nature.[58] Fr. Ashley had a friend from New York who was amazed to learn that the moon rises and sets, since he had

---

[56] Ashley 1995, *passim*.

[57] See Nick Bostrom 2003: "The Transhumanist FAQ — A General Introduction; Version 2.1", exploring "possibilities for the 'posthuman future' created by increased merging of people and technology via bioengineering, cybernetics, nanotechnologies"; frankly secularist.

[58] The current debate whether gender pertains to the essence of marriage is a vivid example.

always seen it shining between high buildings! The goal of human life, as Aristotle argued, is contemplative, and the ultimate value of the scientific exploration of our universe is not to control nature's powers but to wonder at its marvels and at the Creator that nature's powers reveal by analogy from effect to First Cause. On vacation, people naturally incline to leave the technologized city for the country, the forest, and sea-shore of natural beauty.

We can conclude therefore that the Holy Spirit is calling us to a simpler life and to technological inventions that respect nature. Jesus in his preaching constantly urges us to be awake to the fact that this world is going to come to an end in its present form and be transformed in eternity. Therefore, to be *too* absorbed in trying to control and perfect this world is to waste our limited energies. We cannot, however, conclude from this that we no longer have a responsibility "to cultivate" our planet. Jesus taught us to pray in the Holy Spirit, through himself as Divine Son, to our Father: "Thy kingdom come, on earth as it is in heaven". Hence we need to make a good use of the creative gifts God has given us, but without wasting time doing so when we might better employ those gifts in preparation for the next world. This holds for the human race midway between the material and the spiritual regions of creation. The angels may be in "aeveternity", and in their first act entered into the beatific vision or decisively excluded themselves from that vision.[59] But the physical portion of the universe that God

---

[59] Because the concepts formed by the angels are comprehensive rather than discursive, and changes on the earth are governed by the unchanging rotation of the celestial spheres as the "ruling cause" (*causa regitiva*) preventing chance events from being a developmental factor, St. Thomas argued that the original rebellion of Lucifer and his angelic followers was an irreversible choice. But once the unchanging universe has been replaced by an evolutionary one, the angel, turning its attention here or there, is constantly liable to surprises further revealing the limited or finite nature of its intellectual power, for all its "comprehensiveness" at any given moment; and this circumstance introduces into the "angelic world", as it were, a kind of *successive discourse* which might provide over billions of years a sufficient

has so freely and generously chosen to create is in *time*, which He has willed sometime to have an end.

It is often supposed that the Catholic Church teaches that the present world will end in vast disasters just before the Last Judgment. Conservative Protestants take literally St. Paul's words about the "rapture" — "Then we who are alive, who are left, will be caught up together with them in the clouds to meet the Lord in the air" (1 Tm 4:17) — and also the vivid pictures of the End Time in *Revelations*. This overlooks the fact that the Lord taught us to pray "Thy kingdom come ... on earth as it is in heaven" (Mt 6:9–10),[60] and in many passages Jesus speaks of how all nations will come to Jerusalem and worship the true God, etc.

In our opinion, therefore, in view of human freedom by which we can reject God's grace, but are also free by his grace (but *only* by his grace) *fully* to accept and cooperate with his will, these biblical texts should be understood as picturing the alternative histories between which humanity inevitably chooses. When the Lord Jesus returns, he may find the whole of humanity like the prudent virgins: ready for the wedding with lamps burning in their hands; or he may find all with

---

"surprise factor" that even the "fallen angels" might come to realize the "error of their ways". On some such grounds there might be room for the opinion of some modern theologians that even Lucifer might be redeemed in the end!

[60] In Ashley 2009: *Meditations on the Mysteries of Light in the Rosary*, pp. 42–53, is an analysis of the literary structure of the Sermon on the Mount, based on the view of some scholars that it is chiastic (cross-like) in form, climaxing with "So be perfect, just as your heavenly Father is perfect" (Mt 5:48), and then paralleling the first half of the Sermon in reverse order, concluding with the Parable of the Two Houses parallel to the opening Beatitudes. Furthermore, the ending of the first part of the Lord's prayer, "on earth as it is in heaven", applies not only to the third petition, "Thy will be done", but also to the first two petitions. This threefold petition renders the prayer implicitly *Trinitarian*, since "Hallowed be thy Name" fits God the Father as Creator, "Thy Kingdom come" fits the Son who came to preach the Kingdom, and "Thy Will be done" fits the Holy Spirit who, like an act of love, proceeds as an act of will from the Father through the Son (intellect).

their lamps burnt out except for a few, maybe even only a tiny few, with their lamps still aflame. The *Catechism of the Catholic Church* #677 does insist that "The kingdom will be fulfilled, then, not by a historic triumph of the Church through a progressive ascendancy, but only by God's victory over the final unleashing of evil which will cause his Bride to come down from heaven". Although this statement excludes an end to this world in which the Church in its humanity can claim victory by its own power rather than be lifted up to God, it does not, as far as we can see, deny that the Church's human efforts can result in global spread of the Church and in great achievements in social justice and proper use and cultivation of the environment.

The reason we emphasize this historic openness to human freedom is that too often the words of the Lord in praise of the penitent woman who anointed his feet with precious ointment (Mt 26:11: "The poor you will always have with you; but you will not always have me") is treated as an excuse for neglecting the poor! In fact it should be understood to mean "You will always have opportunities to exercise charity, as this woman is doing to me because my death is close at hand". This is why Jesus repeatedly warned us to be ready at all times, because the end of history, like the end of our own lives, has not been revealed to us in anything but a symbolic form, in order that we should act from faith, hope, and love rather than from presumption.

How to Interpret the "Book of Revelation"?

The symbolic language of the Bible clearly predicts "a new heaven and a new earth" (Rev 21:1–5). It is central to Christian tradition as historical that, just as Jesus rose from the empty tomb, so all the dead will be resurrected and regain their individual bodies that will thence exist forever either in Hell or Heaven. Because of the difficulty of fitting this final state of the universe into the modern picture of a Big Bang universe expanding into inevitable entropic doom, modern theologians hesitate to interpret this Biblical language.[61]

The medieval theologians, however, did not hesitate to speculate on this subject in terms of the Platonic/Aristotelian universe with its celestial spheres, as Ptolemy (c.90–168 AD) would formally mathematize as the Greek era of philosphy's beginning neared its end.[62] St. Thomas Aquinas died before completing the part of the *Summa Theologiae* dealing with the "Last Things", but his editors, in the *Supplement* that they fashioned to complete the *Summa* Part III, qq. 69–99,[63] did so very fully using material from his earlier (c.1254/56) *Commentary on the Sentences of Peter Lombard*, IV, dd. 43–50. There Aquinas had entertained the view that, before the Judgment whose time is hidden from us, the region within the sphere of the moon will be consumed by fire, except for the part that will become the Hell of the damned, and for "Limbo" where unbaptized infants will remain in a state of merely natural happiness. Both good and bad humans will be resurrected in their mature bodies, male and female. The damned will retain all bodily defects and suffer both spiritually and physically from their sinful state. The blessed will be freed from Purgatory, a place not primarily of punishment but of purification from the remaining effects of sin, and will be joined with the saints in the eternal beatific vision of God. This beatific vision will also glorify their bodies, so that they will be unable to suffer and will be "subtle, agile, and clear" — that is, although their bodies remain tangible, they will be entirely subject to intelligence and will, able to move anywhere freely, and will shine with degrees of light manifesting their inner virtues. They will have the bodily

---

[61] On current views, see Joseph Ratzinger (now Pope Benedict XVI) 2007: *Eschatology, Death and Eternal Life.*

[62] See Robert Turnbull 1998: 186f.; cited in Deely 2001: 60 note 25.

[63] Consult online <http://www.ccel.org/a/aquinas/summa/XP/XP069.html>; for the Latin, see <http://www.archive.org/stream/operaomniaius sui12thom>.

functions of physical life, but will not eat (nor have need hence of eliminatory functions — see pp. 160–161 below).

These medieval speculations do not consider if the blessed will breathe or have a beating heart. As for the celestial spheres — sun, moon, planets and stars — they will remain as they are, but will cease their circular movement, so that universal time will cease. The sublunar region and the earth will be transformed, so that earth will be a Holy City of the saints, but now without plants or brute animals. The "many mansions" (Jn 14:2) of the saints will simply be their various degrees of beholding the beatific vision. They will move at will about the earth to see and visit with their blessed companions in eternal existence, and the blessed will at the center of their souls see God as He is, One God in Three Persons. Also they will see themselves as perfected in God's image, as Jesus was and is, and as is also already his Blessed Mother, assumed into Heaven with him. Conversely, those who by their own choice have preferred themselves to God will continue to exist bodily but forever in a state of internal contradiction to their nature and to an intimate relation to God, the relation to which they were lovingly called by Him but by their own choices refused.

This graphic picture from theology past no doubt more than warrants the hesitation of theologians today to give a literal interpretation to the "Book of Revelation". In Catholic doctrine as developed today, the view is that in the resurrection the spiritual soul will receive not just some body but the same body that it had at the moment of its conception and when the spiritual soul was infused thereto, although this body will be transformed.[64] It will be "glorified" as was Jesus' body after the resurrection, that is to say, freed of its physical limitations and dependency upon interactions with a material environment. When some of the Corinthian Christians doubted this, St. Paul wrote them as follows (I Cor 15: 35–57):

---

[64] *Catechism of the Catholic Church* 2000: #s 988-1013.

But someone may say, "How are the dead raised? With what kind of body will they come back?" You fool! What you sow is not brought to life unless it dies. And what you sow is not the body that is to be but a bare kernel of wheat, perhaps, or of some other kind; but God gives it a body as he chooses, and to each of the seeds its own body. Not all flesh is the same, but there is one kind for human beings, another kind of flesh for animals, another kind of flesh for birds, and another for fish. There are both heavenly bodies and earthly bodies, but the brightness of the heavenly is one kind and that of the earthly another. The brightness of the sun is one kind, the brightness of the moon another, and the brightness of the stars another. For star differs from star in brightness. So also is the resurrection of the dead. It is sown corruptible; it is raised incorruptible. It is sown dishonorable; it is raised glorious. It is sown weak; it is raised strong. It is sown a natural body; it is raised a spiritual body. If there is a natural body, there is also a spiritual one. So, too, it is written, "The first man, Adam, became a living being", the last Adam a life-giving spirit. But the spiritual was not first; rather the natural and then the spiritual. The first man was from the earth, earthly; the second man, from heaven. As was the earthly one, so also are the earthly, and as is the heavenly one, so also are the heavenly. Just as we have borne the image of the earthly one, we shall also bear the image of the heavenly one. This I declare, brothers: flesh and blood cannot inherit the kingdom of God, nor does corruption inherit incorruption. Behold, I tell you a mystery. We shall not all fall asleep, but we will all be changed, in an instant, in the blink of an eye, at the last trumpet. For the trumpet will sound, the dead will be raised incorruptible, and we shall be changed. For that which is corruptible must clothe itself with incorruptibility, and that which is mortal must clothe itself with immortality. And when this which is corruptible clothes itself with incorruptibility and this which is mortal clothes itself with immortality, then the word that is written shall come about: "Death is swallowed up in victory. Where, O Death, is your victory? Where, O death, is your sting?" The sting of death is sin, and the power of sin is the law. But thanks be to God who gives us the victory through our Lord Jesus Christ.

The Current End Time as Projected by Science

Current science seems to contradict all this flatly, since, as we have already said at the beginning of this Chapter, science no longer holds for a Steady-State Universe, as did Aristotle and Aquinas and 20th century science before the Big Bang theory. The common view currently, as already described, is that the universe will eventually come to a "heat-death" when it reaches a state of maximum entropy, that is, when all the energy within it in the galaxies of stars etc. has dispersed to colder regions, so that no more work of one body on another can take place. This is not identical with "cold death" ("Big Chill") resulting from the Big Bang expansion that might have a reverse into a "Big Crunch". Also, some think it will end in a "Big Rip", in which everybody is blown apart by Dark Energy. Of these hypotheses, however, the dispersal view seems most probable.

In any case, both entropic dispersal and cosmic expansion lead to a dispersal of matter and energy to the point that nothing but chance quantum fluctuations in the void remain. Thus this dispersal results in an "equilibrium", or a certain kind of "order", but since it is an order of purely *random* motion producing no stable results, it is better called a "chaotic disorder". The rate of increase of entropy in the universe as a finite whole, however, depends on many factors, and in limited regions it permits a reversal or increased order and complexity (*negentropy*)[65] — otherwise, the evolution of life in the direction of greater complexity would have been impossible.[66]

---

[65] See the "Negentropy" entry in *Wikipedia* <http://en.wikipedia.org/wiki/Negentropy>.

[66] See Paul Davies 1997: *The Last Three Minutes: Conjectures about the Final Fate of the Universe.*

Our 13.7 billion year aged universe is in its Stelliferous or star-filled time, that is probably now about half over. Then the Degenerate Era will begin as the stars burn up their nuclear fuel and die. The sky will become dark and most material will be locked up in dead stars, white dwarf stars, or black holes. Then comes the era of proton decay, beginning at about 100, trillion, trillion, trillion years from now, and all matter will be sucked into dark holes. Eventually these will also disperse, leaving a sea of electrons, positrons, neutrinos, so that the universe will be totally black at about 1 followed by 200 zeroes.[67] This hardly sounds like the Bible's radiant picture of the end of time (Rev 21:1–5):

> Then I saw a new heaven and a new earth. The former heaven and the former earth had passed away, and the sea was no more. I also saw the holy city, a new Jerusalem, coming down out of heaven from God, prepared as a bride adorned for her husband. I heard a loud voice from the throne saying, "Behold, God's dwelling is with the human race. He will dwell with them and they will be his people and God himself will always be with them (as their God). He will wipe every tear from their eyes, and there shall be no more death or mourning, wailing or pain, (for) the old order has passed away". The one who sat on the throne said, "Behold, I make all things new". Then he said, "Write these words down, for they are trustworthy and true".

Where is the universe going? How will it end? According to the Second Law of Thermodynamics, every system of interacting bodies taken as a whole in the course of time grows more and more random in its actions, although the rate of this decline in order is not specified. Furthermore, there can be temporary reverses toward greater order in regions of the system, though they will eventually smooth out. Hence evolution

---

[67] Michael D. Lemonick 2001: "How the Universe Will End".

toward greater order can occur here or there in the universe as it has on our planet with the rise of life and then the evolution of intelligent life into the human species.

It would appear, therefore, that eventually life will be extinct, on our planet and also on other planets. This prediction is supported by the fact that stars eventually burn up the hydrogen in their cores into helium. Our sun in about 5 billion years will reach this state, and will expand and probably engulf the earth. Eventually all the stars will burn out, and probably only dark matter[68] will remain.

Thus, the Big Bang theory holds that the universe we know is expanding and, due to Dark Energy,[69] at an accelerating speed. Finally it will be so thinned out that no activity will be left except minute quantum fluctuations in a "vacuum", that is, in a field devoid of massive matter; but since it has dimensive quantity it will still be material. In an alternative hypothesis, a Big Crunch would then begin, in which all matter would finally again condense to a "singularity" like that before the Big Bang. If that happens, no effects of the present universe will remain in the new cycle, so that the new cycle cannot be predicted from the history of our cycle, although it may be a simple repetition of it.

Reconciling the Two? (or: Guessing the Unguessable)

Yet the history of the universe *as a whole* may perhaps go on to this predicted condition of maximum entropy and mere quantum fluctuations in a void, while some region of it is preserved at a higher level of order. Somewhat as the Many Worlds Theory hypothesizes "bubble universes" within a super-universe, the new heaven and new earth may be a region which at the Judgment will be miraculously saved forever from the running

---

[68] See the "Dark Matter" *Wikipedia* entry <http://en.wikipedia.org/wiki/Dark_matter>.

[69] See the "Dark Energy" *Wikipedia* entry <http://en.wikipedia.org/wiki/Dark_energy>

down of the present universe as a whole, just as in evolutionary theory the earth has been for a time a region of "negative
entropy", that is, of increasing, not declining, order. Thus, the
new heaven and new earth will arise *within* that increasing des-
ert, as Eden was a garden in a waste land (Gn 2:4-6):

> At the time when the LORD God made the earth and the
> heavens — while as yet there was no field shrub on earth
> and no grass of the field had sprouted, for the LORD
> God had sent no rain upon the earth and there was no
> man to till the soil, but a stream was welling up out of the
> earth and was watering all the surface of the ground —

Inasmuch as present scientific knowledge tolerates a Multiverse hypothesis, this state of affairs would not contradict anything that we now know.

Earlier, we admitted the possibility of extraterrestrial intelligent species, each with its own history. Some may have
been created in a state of pure nature, while perhaps others
were created in the state of grace as we humans are thought to
have been. What is certain is that those of our world admitted to the beatific vision are redeemed through Jesus Christ,
keeping in mind (as Aquinas points out[70]) that the Son of God
— the Second Person of the Trinity — may have been in-
carnated in different bodies on different worlds. It is perhaps

---

[70] Aquinas 1266/73: *Summa theologiae* III, q. 3, art. 7: "What has power
for one thing, and no more, has a power limited to one. Now the power of a
Divine Person is infinite, nor can it be limited by any created thing. Hence it
may not be said that a Divine Person so assumed one human nature as to be
unable to assume another. For it would seem to follow from this that the Personality of the Divine Nature was so comprehended by one human nature as
to be unable to assume another to its Personality; and this is impossible, for
the Uncreated cannot be exhausted by any creature. Hence it is plain that,
whether we consider the Divine Person in regard to His power, which is the
principle of the union, or in regard to His Person, which is the term of the
union, it has to be said that the Divine Person, over and beyond the human
nature which He has assumed, can assume another distinct human nature."
See the discussion in O'Meara 1997: *Thomas Aquinas, Theologian.*

possible, but also quite unlikely (from what we know of the limitations of animal life, even when endowed with the intelligence and freedom consequent upon an infused soul), that on some planets embodied intelligences were created in a state of pure nature and remained sinless![71] So perhaps some ET

---

[71] So, returning to the question as discussed on pp. 116–118 above, "Does original sin exist or not?", Pope Benedict XVI (2008: pars. 3–7) answers as follows: "we must distinguish between two aspects of the doctrine on original sin. There exists an empirical aspect, that is, a reality that is concrete, visible, I would say tangible to all. And an aspect of mystery concerning the ontological foundation of this event. The empirical fact is that a contradiction exists in our being. On the one hand every person knows that he must do good and intimately wants to do it. Yet at the same time he also feels the other impulse to do the contrary, to follow the path of selfishness and violence, to do only what pleases him, while also knowing that in this way he is acting against the good, against God and against his neighbour. In his Letter to the Romans St Paul expressed this contradiction in our being in this way: 'I can will what is right, but I cannot do it. For I do not do the good I want, but I do the evil I do not want' (7: 18–19). This inner contradiction of our being is not a theory. Each one of us experiences it every day.

"As a consequence of this evil power in our souls, a murky river developed in history which poisons the geography of human history. ...

"Thus, the existence of the power of evil in the human heart and in human history is an undeniable fact. The question is: how can this evil be explained? In the history of thought, Christian faith aside, there exists a key explanation of this duality, with different variations. This model says: being in itself is contradictory ... being as such bears within itself both evil and good from the outset. Being itself is not simply good, but open to good and to evil ... equally primal with the good. ... What Christians call original sin would in reality be merely the mixed nature of being. ...

"And let us therefore ask again: what does faith witnessed to by St Paul tell us? As to the first point, it confirms the reality of the competition between the two natures, the reality of this evil whose shadow weighs on the whole of Creation. ... Quite simply, evil exists. As an explanation ... faith tells us: there exist two mysteries, one of light and one of night. ... The first mystery of light is this: faith tells us that there are not two principles, one good and one evil, but there is only one single principle, God the Creator, and this principle is good, only good, without a shadow of evil. And therefore, being too is not a mixture of good and evil; being as such is good and therefore it is good to be, it is good to live. This is the good news of the faith: only one good source exists, the Creator. ... Then follows a mystery of darkness, or night. [Moral] evil does not come from the source of being

species fell and was redeemed to exist again in a purely natural state; or, as happened on our own planet, a species has been raised to grace and awaits the Judgment; or have all died unrepentant. We simply know nothing of the realization of such possible diverse histories of spiritualized animals, except that in the final state of the universe the Son of God Incarnate will be the supreme goal of the new heaven and earth, and of the community of created intelligent beings that it will contain (including good angels).

Certainly Catholic theology, since it holds for the resurrection of Jesus and assumption of Mary (and does not positively exclude that of other saints), has to suppose that they even now exist elsewhere in the universe. There is plenty of room! The Bible also seems to indicate that these assumed "Holy Ones" will return from where they now are to our earth where they passed their former lives. Certainly Jesus in his resurrected body, along with his Mother who, because of her sinlessness in soul and body, was "assumed into heaven", and possibly other saints such as Elijah, who "went up to heaven in a whirlwind" (2 Sm 2:11), are even now located somewhere in our universe that is appropriate to their glorified state and anticipates the new heaven and earth.

What then of Hell? It is not revealed to us who will be imprisoned in Hell, but it will be a part of the universe in which will be bound those who at death, in whatever situation they died, have still refused God's mercy as offered to them through

---

itself, it is not equally primal. [Moral] Evil comes from a freedom created, from a freedom abused.

"... [Moral] Evil is not logical. Only God and good are logical, are light. [Moral] Evil remains mysterious. ... chapter 3 of Genesis ... makes us guess but cannot explain what is itself illogical. ... It remains a mystery of darkness, of night ... from a subordinate source. ... And therefore evil can be overcome. Thus the creature, man, can be healed. ... God introduced healing. He entered into history in person. He set a source of pure good against the permanent source of evil. The Crucified and Risen Christ, the new Adam, counters the murky river of evil with a river of light. And this river is present in history."

the Holy Spirit. As a work of art with beautiful colors is enhanced by a few dark lines or areas to bring out these colors fully in their beauty, so the darkness of Hell will enhance the glory of the new heaven and the new earth. How many resurrected persons (along with evil angels) will be confined to this eternal prison we have not been told; but we have been mercifully warned against choosing that self-inflicted doom. Some, like Origen (c. 185–254 AD) and the modern theologian Hans Urs von Balthasar (1905–1988), have argued from the mercy and omnipotence of God that probably Hell will be empty; but from the words of Jesus concluding the Sermon on the Mount, this seems unlikely (Mt 7:21–23):

> Not everyone who says to me, "Lord, Lord", will enter the kingdom of heaven, but only the one who does the will of my Father in heaven. Many will say to me on that day, "Lord, Lord, did we not prophesy in your name? Did we not drive out demons in your name? Did we not do mighty deeds in your name?" Then I will declare to them solemnly, "I never knew you. Depart from me, you evildoers".

The transformation of these special regions of the universe need not, in our opinion, be thought of as a direct act of God, since, as previously argued, the angels would be created spirits with some power over the material forces of the universe. Thus it may well be that God will give to human souls powers by which, as instruments of God, they reconstruct their own bodies. Furthermore, the entire universe will be transformed as a fitting home for these glorified bodies, even as by the Anthropic Cosmological Principle it has been made the fitting site for human evolution. This is no more extravagant a surmise than the hopes of the "Transhumanists"[72] to make us immortal by unaided human powers.

---

[72] See the *Wikipedia* entry "Transhumanism" at <http://en.wikipedia.org/wiki/Transhumanism>.

Thus the transformation of this radiant universe within the sea of darkness left by the old universe (Rev 21:1, previously quoted, says that "the sea will be no more") may therefore be accomplished through the community of pure and resurrected spirits acting as God's instruments. In the outer reaches of this special universe would be the realm of limbo for unbaptized infants and of extraterrestrials created in the state of pure nature (if such there be). Still farther out would be the region of the Hell of the damned, but still able to sustain human bodies even after the entropic death of the total universe.

What might that new heaven and earth in which the blessed will dwell be like? The Scriptures call this transformed region the "New Jerusalem" with its limiting walls, and describes it as if it were glowing with light. Dante pictured it as a Rose of Light, but made explicit that this was a symbol of something transcending imagination.[73] The fact that time will have ceased for the Blessed seems to imply that nothing within "heaven" will be in motion, although Fra Angelico pictured the blessed as dancing in circles. The Risen Christ moved at will, passing through solid walls, yet also at will offered his body to the verifying touch of the Apostle Thomas (Jn 20:27). But these were manifestations in time, and seem to have no place in eternity. Thus might it be better to imagine heaven as a globe of light (as if light were the highest of material entities) within a larger vacuous universe. Yet light is a vibration, and so some changeless super-light might be more suitable.

Within this heaven will be preserved all the bodies of the blessed and all the monuments of their history, such as the hill of Calvary, in an unchanging work of art manifesting outwardly their collective internal memories of the past. The meaning of these memories, now so dark and confused, will be present in their souls; but will at last be seen as part of the greater good that God intended to consummate. Thus all former evils will

---

[73] Dante Alighieri 1306–1321: *La Divina Comedia*, Canto 4, 28ff.

be like music reduced to a single comprehensive note, the final chord of a symphony summing up all that went before.

Aquinas viewed the transformed human body as perfectly mature, glorious, agilely mobile, yet needing neither to eat nor to reproduce. He says nothing of its needing to breathe or having a circulation of the blood, and his medical views on such matters were so imperfect that it is just as well. Certainly the Blessed do not need to reproduce or enjoy sex, as the probably allegorical[74] language of the *Qur'an* pictures them doing, because the body will be filled with a greater joy by the overflow of the beatific vision. In the present life the human intellect was created so as to be able to render actually intelligible the things around us,[75] and thus our intellectual activity and free will require no further efficiency (except the First Cause), but emanate from the intellectual soul.

Our bodies, however, now require food and also oxygen from respiration, principally, first, to keep alive the brain, the necessary instrument of the intelligence; and then sent to meet the energy needs of our other organ systems. But it seems possible that this corporeal energy in the risen body will come from the eternal "overflow" of the spiritual intelligence, and will thus render unnecessary food, air, and the circulation of the blood. Surely our sense powers and affective powers will still operate, since we will still experience the physical aspects of joy and sympathy. All these powers are directed not only to

---

[74] Cf. Aquinas i.1259/65: *Summa contra gentiles* Book 1, chap. 6 n. 7.

[75] The reference here is to Aquinas' notion of "*intellectus agens*" whereby we able to form intellectual concepts over and above the concepts (the "phantasms") of animal internal sense. For a full discussion to date, in relation to Aquinas 1266: *Summa theologiae* I, q. 79 "Of the intellectual powers", esp. arts. 3–5, q. 84 "How the soul while united to the body understands corporeal things", esp. arts. 6–7, and q. 85 "Of the mode and order of understanding", esp. arts. 1–3, see the following three texts by Deely: 1971, "Animal Intelligence and Concept-Formation"; 2007a: *Intentionality and Semiotics*, esp. "The Problem of Actual Intelligibility" and "Abstraction", pp. 39–114; and 2007/8. "The Quo/Quod Fallacy in the Discussion of Realism".

maintaining the body in its integrity, which in beatitude will not be necessary, but to the service of its spiritual powers. For example, the blessed may breathe not in order to get energy for the brain, as in this life, but in order to speak and sing as depicted in the Scripture, since they will still communicate with each other and express themselves in a human mode.

In *Theologies of the Body: Humanist and Christian*[76], Ashley suggests that perhaps the "spiritual body", as St. Paul calls it (I Cor 15:35–57, quoted pp. 151–152 above), might be simply the *information* present in the body acquired in conception and by the time of death converted into a field of light waves — a sort of cyberspace reality. Ashley there called this "guesswork", and here reconsiders it to be simply an unnecessary hypothesis. By "spiritual" St. Paul may simply have meant that the glorified body is totally under the control of the spiritual intelligence and will.

Although there will be no universal time in the proper sense (since in the universe as a whole orderly change will supposedly have ceased), it does not seem that this excludes a super-time of endlessly increasing comprehension of the vision of God. No creature, even in the beatific vision, can comprehend the plenitude of God. Hence it seems that although the beatific vision attains to the very essence of God as Trinity, it is possible for it to deepen and forever continue to deepen, while still maintaining a degree of perfection that corresponds to the degree of charity at death.

Furthermore, the eternal conversation among the Blessed will be a sharing between those who have gained a deeper vision and others who have not, and perhaps by each person according to her or his special gifts. We would take this to be the allegorical meaning of the singing of the praises of God, so often depicted in the Scriptures.

---

[76] See Ashley 1996: *Theologies of the Body: Humanist and Christian*, pp. 588–620, where eschatology is dealt with at some length.

Aquinas distinguishes in the natural order the difference of "time" as we humans experience it in a quantitative material universe, and the "aeveternity" of the angels that is only analogous to time but is an order of distinct acts of thought.[77] Perhaps in the same way we should understand that although in the beatific vision "time is no more" (Rev 10:6), this refers to earthly time, yet analogically there is an everlasting advance in the communal vision the blessed have of God.

Therefore, while science cannot rationally predict the theological End Time, its present hypothesis about the heat-death of the universe need not contradict the Biblical view of a new heaven and a new earth as the end and goal of human history. Moreover, the current scientific view supports and enriches the idea by suggesting how our material universe may be in an evolutionary process beginning with the Big Bang and ending with the Big Chill — nothing existing before it, and after it only featureless space. Yet the evolution of the present universe has not been in vain, but has been guided by God to the production of human animals spiritualized in their souls and actions as a marvelous work of art that will be preserved forever in a miniverse of glory suited to the human and angelic community as they gaze on the face of God forever in mutual sharing and deepening love. The great achievement of science will be to help us in our growing understanding of God's plan, an acheivement that will be preserved forever in the minds of the blessed.

---

[77] Aquinas 1266: *Summa theologiae* I, q. 10 art. 5: "the angels ... have an unchangeable being as regards their nature, with changeableness as regards choice; moreover, they have changeableness of intelligence, of affections and of places, in their own degree. Therefore these are measured by aeveternity, which is a mean between eternity and time". Cf. the analysis in Deely 2004: "The Semiosis of Angels". The basic idea is that *time* is the measure of motion in the order of substances which themselves undergo generation and corruption (beginning and ceasing to exist), while *aevum* is the measure of accidental changes in and among substances which cannot undergo corruption (ceasing to exist). See the *Wikipedia* entry "Aevum" (redirected from "Aeveternity").

Thus the Holy Spirit, who is the Divine Plenitude, by the power of the Father and in the likeness of the Son, will have completed the work of bringing the creation to its full perfection in which all creatures have returned to God according to their natures transformed by grace. The Holy Spirit thus fulfills the goal that God the Father intended for his creation, patterning it after his Son, the Word who has entered into our universe in the Incarnation, and completing it as the glorified universe at whose center is his community, the Church, whose members are the beatified saints and angels.

It is the business of science to more and more understand this universe in its complexity and dynamism, and thus to enrich and correct, as may be, the details of the *outline* furnished us by faith and developed by theology. The technological application of scientific knowledge, thus, as we noted above,[78] will helps us "care and cultivate" our "garden earth"; but by comparison with our practical knowledge, the contemplative value of scientific advance in helping us understand the meaning and wonder of our human existence is the far greater.

---

[78] See in this Chapter p. 144f., esp. at note 53.

BIBLIOGRAPHY OF REFERENCES,
HISTORICALLY LAYERED

Sacred Texts

BIBLE, The Holy, citations are from *The New American Bible* (NAB), with the revision of the Book of Psalms and New Testament, <http://www.vatican.va/archive/ENG0839/_INDEX .HTM>, or <http://www.usccb.org/nab/bible/alpha. htm>; in consultation with *The New American Bible, Revised Edition* (NABRE) <http://www.usccb.org/nab/ bible/index.shtml>.

QU'RAN, The Holy, trans. M. Pickthal (translation: 002.087), in the University of Southern California *Compendium of Muslim Texts* at <http://www.usc.edu/dept/MSA/quran/>.

Church Documents

CATECHISM OF THE CATHOLIC CHURCH.
2000. 2nd Edition (Libreria Editrice Vaticana, and Washington, DC: United States Catholic Conference).

VATICAN COUNCIL I (8 DECEMBER 1869–1870 OCTOBER 20).
1870, April 24. *Dei Filius.* Dogmatic Constitution on the Catholic Faith. See <http://www.disf.org/en/documentation/11-VaticanCouncilI.asp> or <http://www.papalencyclicals. net/Councils/ecum20.htm>.

VATICAN COUNCIL II (11 OCTOBER 1962–1965 DECEMBER 8).
1964, November 21. *Lumen Gentium*, Dogmatic constitution on the Church, promulgated by Pope Paul VI, Chapter VIII, n. 53. <http://www.vatican.va/archive/hist_ councils/ii_vatican_council/documents/vat-ii_const_ 19641121_lumen-gentium_en.html>
1965, December 7. *Gaudium et Spes [Joy and Hope]*, Pastoral constitution on the Church in the modern world, promulgated by Pope Paul VI. <http://www.vatican.va/ archive/hist_councils/ii_vatican_council/documents/ vat-ii_const_19651207_gaudium-et-spes_it.html>

BENEDICT XVI, Pope (Joseph Ratzinger, 16 April 1927– ; elected pope 19 April 2005).
2008, December 3. General Audience on "The Apostle's teaching on the relation between Adam and Christ". Online <http://www.vatican.va/holy_father/benedict_xvi/audiences/2008/documents/hf_ben-xvi_aud_20081203_en.html>; also reported online at *Zenit* under the title "On Christ, the New Adam", at <http://www.zenit.org/article-24456?l=english>.
2009. Homily "about the prayer which concludes these Vespers" celebrated with the Faithful of Aosta, Italy, Friday 24 July 2009: <http://www.vatican.va/holy_father/benedict_xvi/homilies/2009/documents/hf_ben-xvi_hom_20090724_vespri-aosta_en.html>.

JOHN PAUL II, Pope (Karol Jósef Wojtyła, 18 May 1920–2005 April 2; reigned as pope 16 October 1978–2005 April 2).
1981. October 1 "Address to the Pontifical Academy of Science", as published in *L'Osservatore Romano* 4 (November 1992), http://www.silk.net/RelEd/sciencejp.htm.
1996. October 22, "Message to the Pontifical Academy of Sciences: on Evolution", published in *L'Osservatore Romano* (October 30); available online at <http://www.newadvent.org/library/docs_jp02tc.htm>, also at <http://www.ewtn.com/library/PAPALDOC/JP961022.HTM>.
1998. Encyclical, *Fides et Ratio*, <http://www.vatican.va/holy_father/john_paul_ii/encyclicals/documents/hf_jp-ii_enc_15101998_fides-et-ratio_en.html>.
1999. Congregation for the Doctrine of the Faith, "Notification about the book *Jesus Symbol of God* [Maryknoll, NY: Orbis Books, 1999] by Fr. Roger Haight, S.J." *L'Osservatore Romano*, Vatican, Roger Haight Feb. 7–8, 2005 http://ncronline.org/mainpage/specialdocuments/haight_notice.pdf

PIUS XII, Pope (born 2 March 1876, Giovanni Pacelli; elected Pope 2 March 1939–1958 October 9).
1947. Apostolic Constitution, *Provida Mater*, 2 Febr. 1947: AAS 39 (1947) p. 120 ss. <http://www.vatican.va/holy_father/pius_xii/apost_constitutions/documents/hf_p-xii_apc_19470202_provida-mater-ecclesia_it.html>

1950 August 12. Encyclical *Humani Generis*. <http://www
.vatican.va/holy_father/pius_xii/encyclicals/documents/
hf_p-xii_enc_12081950_humani-generis_en.html>

1950 December 8. Allocution *Annus sacer* aux membres du pre-
mier Congrès international des religieux, à Rome. AAS
XLIII (1951) 29.

1951 November 22. "The Proofs For The Existence Of God In
The Light Of Modern Natural Science", Address to the
Pontifical Academy of Sciences.

1960. Posthumously reported *Monitum* address to the Pontifical Academy of Sciences; cf. *Ephemerides Mariologicæ* 11
(1961) 137–138.

Books and Articles (including web citations)

ADLER, Mortimer J. (28 December 1902–2001 June 28).
1967. *The Difference of Man and the Difference It Makes* (New
York: Holt, Rinehart & Winston).
1993. *The Angels and Us* (New York: Collier Books; Toronto,
Canada: Maxwell Macmillan).

ALLEN, John L.
2009. "Pope cites Teilhardian vision of the cosmos as a 'living
host'," *National Catholic Register* (28 July 2009), article
with 84 comments.

AQUINAS, Thomas.
i.1252–1273. *S. Thomae Aquinatis Opera Omnia ut sunt in indice
thomistico*, ed. Roberto Busa (Stuttgart-Bad Cannstatt:
Frommann-Holzboog, 1980), in septem volumina:
1. In quattuor libros Sententiarum;
2. Summa contra Gentiles, Autographi Deleta, Summa Theologiae;
3. Quaestiones Disputatae, Quaestiones Quodlibet-
ales, Opuscula;
4. Commentaria in Aristotelem et alios;
5. Commentaria in Scripturas;
6. Reportationes, Opuscula dubiae authenticitatis;
7. Aliorum Medii Aevi Auctorum Scripta 61.
c.1252/56. *De Ente et Essentia*, in Busa ed. vol. 3, 583–587. This
early work is in effect the commentary of Aquinas on

Porphyry AD 271. English trans. Armand Maurer, *On Being and Essence* (Toronto: Pontifical Institute, 1968); trans. with commentary by Joseph Bobik, Aquinas on *Being and Essence. A translation and interpretation* (Notre Dame, IN: University of Notre Dame Press, 1965).

c.1254/56. *In quattuor libros sententiarum Petri Lombardi*, in Busa ed. vol. 1.

1259/65. *Summa Contra Gentiles* (4 books, often referred to simply as *Contra Gentes*; in Busa ed. vol. 4, 539–542), English title *On the Truth of the Catholic Faith* (Garden City, NY: Doubleday & Co., 1955), Book I trans. Anton C. Pegis, Book II trans. James F. Anderson, Book III trans. Vernon J. Bourke, Book IV, trans. Charles J. O'Neil.

c.1264. "On Reasons for Religious Belief" (*De Rationibus Fidei*, in Busa vol. 3, 509–513). English trans. with a "Foreword" by Joseph Kenny, OP, with the full title "Reasons for the Faith against Muslim Objections (and one objection of the Greeks and Armenians) to the Cantor of Antioch" [*De rationibus fidei contra Saracenos, Graecos et Armenos, ad Cantorem Antiochiae*), published in *Islamochristiana* 22 (Rome, 1996), 31–52.

c.1266/73. *Summa theologiae*, in Busa ed. vol. 2, 184–926.

i.1267/68. *Quaestio disputata De spiritualibus creaturis*, in Busa ed. vol. 3, 352–368.

1271. "On the Eternity of the World" (*De aeternitate mundi*), in Busa ed. vol. 3, p. 591.

1271a. *De substantiis separatis*, in Busa ed. vol. 3, pp. 515–525.

c.1272/73. *In libros de coelo et mundo*, in Busa ed. vol. 4, 1–49.

ARISTOTLE (384–322BC).

Note: our citations here are from the 12-volume Oxford edition (at the Clarendon Press) prepared under W. D. Ross Ed. 1928–1952. For the convenience of the reader, after the abbreviation RM, we also give the pages where applicable to the more readily available one-volume edition of *The Basic Works of Aristotle* prepared by Richard McKeon, using the Oxford translations (New York: Basic Books, 1941). Chronology for the works is based on Gauthier 1970, condensed as follows:

c.330 BC. *On the Soul* (*De Anima*, trans. J. A. Smith; RM 533–603 complete).

c.335/34 BC. *Poetics* (Oxford Vol. XI, 1447a–1462b20, trans. Ingram Bywater; RM 1455–1487 complete).

i.345/4–330 BC. *On the Parts of Animals*, in 4 Books (Oxford Vol. V, 639a1–697b31, trans. William Ogle; RM 643–661, only Chs. 1–5 of Book I, and Ch. 1 of Book II).

i.348–330 BC. *Metaphysics*, Books I–XIV (Ross trans. in Oxford Vol. VIII, 980a1–1093b29; in RM 681–926).

i.353–347 BC. *Physics*, Books I–VIII (Hardie and Gaye trans. in Oxford Vol. II, 184a9–267b26; in RM 213–394).

c.355BC. *On the Heavens* (*De Caelo*) (trans. Harold J. L. Stocks; RM 395–466, missing chs. 1–12 of Book II).

ARMSTRONG, Dave.

2004. "Early Protestant Hostility Towards Science" (Revised and expanded 09 July 2004); <http://socrates58.blogspot.com/2004/07/early-protestant-hostility-towards.html>.

ASHLEY, Benedict M., OP.

1965. "Significance of Non-Objective Art", *Proceedings of the American Catholic Philosophical Association* 39, pp. 156–165.

1973. "Change and Process", in *The Problem of Evolution*, ed. John N. Deely and Raymond J. Nogar (New York: Appleton-Century-Crofts), pp. 265–294.

1995. *Theologies of the Body: Humanist and Christian* (2nd printing; Braintree, MA: Pope John XXIII Medical-Moral Research and Education Center , now National Catholic Bioethics Center, Philadelphia/Boston).

2000. *Choosing a Worldview and Value System: An Ecumenical Apologetics* (Staten Island, NY: Alba House).

2006. *The Way Toward Wisdom: An Interdisciplinary and Intercultural Introduction to Metaphysics* (Notre Dame, IN: University of Notre Dame Press).

2006a. *The Ashley Reader: Redeeming Reason* (Naples, Fl: Sapientia Press).

2009. *Meditations on the Mysteries of Light in the Rosary* (Staten Island, NY: St. Paul's/Alba House).

2011?   *Healing for Freedom: A Christian Perspective on Personhood and Psychotherapy* (Moncton, New Brunswick, Canada: Atlantic Baptist University, forthcoming).

2011a?   *A Marian Ecclesiology* (forthcoming).

ATHEISM, Positive.

2008   "Positive Atheism's Big List of Steven Weinberg Quotat ions" at <http://www.positiveatheism.org/hist/quotes/weinberg.htm>.

ATKINSON, Nancy.

2008.   "Are the Laws of Nature the Same Everywhere in the Universe?", *Universe Today* (20 June 2008); online at <http://www.universetoday.com/15220/are-the-laws-of-nature-the-same-everywhere-in-the-universe/>.

AZAR, Beth.

1999.   "Researchers counting on animals for clues to math", *Monitor OnLine* 30.4 (April 1999; <http://www.apa.org/monitor/apr99/math.html>); available also via <http://www.sciencenewsforkids.org/articles/20031008/refs.asp>.

BAKAN, David.

1958.   *Sigmund Freud and the Jewish Mystical Tradition* (Princeton, NJ: Van Nostrand; reprinted Boston: Beacon Press, 1975).

BARBOUR, Ian.

2002.   *When Science Meets Religion: Enemies, Strangers, or Partners?* (San Francisco: Harper).

BARROW, John D., and Frank J. TIPLER.

1986.   *The Anthropic Cosmological Principle* (Oxford, UK: Clarendon Press).

BEHE, Michael.

2010.   Web Page with answers to critics and bibliography <www.arn.org/authors/behe.html> (part of the "Access Research Network" <http://www.arn.org/>).

BERGSON, Henri.

1907.   *L'Évolution créatrice* (Paris: Librairies Félix Alcan et Guillaumin Réunies). Authorized English trans. by Arthur Mitchell, with a "Foreword" by Irwin Edman, *Creative*

Evolution (New York: Henry Holt & Company, 1911; re-printed New York: Modern Library, 1941).

BERMAN, Jerry.
2004.   "Was Charles Darwin Psychotic? A Study of His Mental Health", *Impact* #367 (January 2004), online at <http://www.icr.org/i/pdf/imp/imp-367.pdf>; also <http://www.icr.org/article/112/8/> or <www.icr.org/index.php?module=articles&action=view&ID=112-25k>.
2010a. "Bertrand Russell" *Stanford Encyclopedia of Philosophy* <http://plato.stanford.edu/entries/russell/>.
2010b. "Russells' Paradox" *Stanford Encyclopedia of Philosophy* <http://plato.stanford.edu/entries/russell-paradox/>.

BESSEL, Friedrich.
2010.   "Biography" in *MacTutor History of Mathematics*, on-line at <http://www-groups.dcs.st-and.ac.uk/~history/BiogIndex.html>.

BLACKWELL, Richard.
1961.   "The Structure of Wolffian Philosophy", *The Modern Schoolman* 38, 303–318.
1992.   *Galileo, Bellarmine, and the Bible* (Notre Dame, IN: University of Notre Dame Press).

BLOOM, Harold.
2006.   "The Heretic Jew", *New York Times Book Review* (18 June 2006), 7 (review of Goldstein 2006).

BOESCH, Christophe, and Michael TOMASELLO.
1998.   "Chimpanzee and Human Cultures", with peer commentary and authors' reply, *Current Anthropology* 39.5 (December 1998), 591ff. Online at <http://cogweb.ucla.edu/Abstracts/Boesch_Tomasello_98.html>.

BONNETTE, Dennis.
2003.   *Origin of the Human Species* (2nd ed.; Ypsilanti, MI: Sapientia Press).

BOSTROM, Nick.
2003.   "The Transhumanist FAQ — A General Introduction; Version 2.1", at <http://www.transhumanism.org/resources/FAQv21.pdf>.

BRITT, Robert Roy.
    2004.   "New study super-sizes the universe: Width of cosmos
            estimated at 156 billion light-years", *Space* on MSNBC.
            com (24 May 2004); <http://www.msnbc.msn.com/
            id/5051818/ns/technology_and_science-space/>.

BUSA, S.J., Padre Roberto, Editor.
    1980.   *S. Thomae Aquinatis Opera Omnia*: see under AQUINAS,
            Thomas, i.1252–1273.

CALKINS, Monsignor Arthur B.
    2005.   "The *Virginitas in Partu* Revisited", online at <http://
            www.airmaria.com/vlog/stnd/stnd0002MsgrCalkins3
            .asp>.
    2006.   "Our Lady's Virginity in Giving Birth", online at <http://
            www.motherofallpeoples.com/Articles/General_
            Mariology/our-ladys-virginity-in-giving-birth.html>

CHOMSKY, Noam.
    1966.   *Cartesian Linguistics: A Chapter in the History of Rationalist
            Thought* (New York: Harper & Row).

CLOONEY, Francis X., SJ.
    2005.   *Divine Mother, Blessed Mother* (New York, NY: Oxford
            University Press).

COBB, John B., Jr.
    1975.   *Christ in a Pluralistic Age* (Philadelphia: Westminster
            Press, 1975). See the Excerpt prepared for "Religion On-
            line" by Harry and Grace Adams: <http://www.religion
            -online.org/showchapter.asp?title=403&C=128>.

COBLEY, Paul, Editor.
    2009.   *Realism for the 21st Century. A John Deely Reader* (Scran-
            ton, PA & London, UK: University of Scranton Press).

DANIÉLOU, Jean, Cardinal.
    1964.   *The Theology of Jewish-Christianity* (Volume 1 of *The De-
            velopment of Christian Doctrine Before the Council of Nicea*),
            trans. and ed. John A. Baker (Chicago: H. Regnery).

DANTE ALIGHIERI.
    1306–1321. *La Divina Comedia*, in the Allen Mandelbaum trans.
            with facing Italian original, *Divine Comedy* (Berkeley,
            CA: University of California Press, 1980, 1982, 1984).

DARWIN, Charles.
  1860. Letter to Asa Gray dated 22 May, #2814 online at
         <http://www.darwinproject.ac.uk/entry-2814> from *The
         Correspondence of Charles Darwin*, published in chrono-
         logical order, ed. F. Burkhardt *et al.* (18 volumes as of
         2010; Cambridge, UK: University of Cambridge Press,
         1985–?), Volume 8, 1860, ed. Frederick Burkhardt, Janet
         Browne, Duncan M. Porter, Marsha Richmond (Cam-
         bridge, UK: Cambridge University Press, 1993). See
         "Darwin Correspondence Project" <http://www.darwin
         project.ac.uk/home>.

DAVIES, Paul.
  1997. *The Last Three Minutes: Conjectures about the Final Fate of
         the Universe* (new edition; New York: Basic Books).

DAWKINS, Richard.
  2006. *The God Delusion* (Boston, MA: Houghton Mifflin).

DE FINANCE, Joseph.
  1955. *Existence et Liberté* (Paris: Vitte).

DE KONINCK, Charles (29 July 1906–1965 February 13).
  1935. *Le Problème de l'Indéterminisme* (Quebec, Canada: Acad-
         emie Canadienne Saint-Thomas d'Aquin, sixième ses-
         sion), 65–159.
  1937. "Réflexions sur le Problème de l'Indéterminisme", *Re-
         vue Thomiste* XLIII Nos. 2 & 3, pp. 227–252 & 393–409.

DE LUBAC, Henri.
  1946. *Surnaturel* (Paris: Aubier, 1946).
  1967. *The Mystery of the Supernatural*, trans. Rosemary Sheed
         (New York: Herder and Herder).
  1969. *Augustinianism and Modern Theology*, trans. Lancelot
         Sheppard (London: Chapman, 1969).
         This translation, like the original, has numerous
         untranslated Latin citations. In new edition, adding
         an "Introduction" by Louis Dupré (New York: Cross-
         roads Publishing Co., 2000), reviewer Aidan Nichols,
         OP (2001: 88), points out that the publisher "wisely put
         into English for this new edition the chunks of patris-
         tic, sixteenth-century, and later Latin which previously
         barred the unwary reader's road. Regrettably, however,

the translations thus provided leave on occasion a good deal to be desired" by reason of the "imperfect comprehension of the thought world to be found in de Lubac's sources." Although the new 2000 reprint "usefully (especially for the checking of the Latin translations) includes in its margins the pagination of the French original", the reader interested in the original Latin, especially knowing that it has not been reliably translated throughout, will need to have as well the earlier edition for the sake of consulting the actual Latin texts.

1984. *A Brief Catechesis on Nature and Grace*, trans. Richard Arnandez (San Francisco: Ignatius Press).

DEELY, John.

1965/66. "Evolution: Concept and Content", *Listening*, Part I in Vol. 0, No. 0 (Autumn), 27–50; Part II in Vol. 1, No. 1 (Winter), 35–66.

1966. "The Emergence of Man: An Inquiry into the Operation of Natural Selection in the Making of Man", *The New Scholasticism* XL.2 (April), 141–176.

1966a. "The Vision of Man in Teilhard de Chardin", *Listening* Vol. 1, No. 3 (Autumn), 201–209.

1969. "The Philosophical Dimensions of the Origin of Species", *The Thomist* XXXIII (January and April), Part I, 75–149, Part II, 251–342.

1969a. "Evolution and Ethics", *American Catholic Philosophical Association Proceedings* XLIII, 171–184; reprinted in Cobley Ed. 2009: 54–73.

1971. "Animal Intelligence and Concept-Formation", *The Thomist* XXXV.1 (January 1971), 43–93; reprinted in Cobley Ed. 2009: 91–139.

1980. "The Nonverbal Inlay in Linguistic Communication", in *The Signifying Animal*, ed. Irmengard Rauch and Gerald F. Carr (Bloomington: Indiana University Press), 201–217.

1982. *Introducing Semiotic. Its History and Doctrine* (Bloomington, IN: Indiana University Press).

1986. "Doctrine", terminological entry for the *Encyclopedic Dictionary of Semiotics*, ed. Thomas A. Sebeok et al. (Berlin: Mouton de Gruyter), Tome I, p. 214.

1994. *The Human Use of Signs; or Elements of Anthroposemiosis* (Lanham, MD: Rowman & Littlefield).

1997. "The Seven Deadly Sins and the Catholic Church", *Semiotica* 117–2/4, 67–102.

2001. *Four Ages of Understanding: The First Postmodern Survey of Philosophy from Ancient Times to the Turn of the Twentieth Century* (Toronto: University of Toronto), pp. 605, 622; and see the Index entry "semiotic web".

2002. *What Distinguishes Human Understanding?* (South Bend, IN: St. Augustine's Press).

2004. "The Semiosis of Angels", *The Thomist* 68.2 (April 2004), 205–258.

2004a. "Dramatic Reading in Three Voices: 'A Sign Is *What?*',", *The American Journal of Semiotics* 20.1–4, 1–66.

2006. "The literal, the metaphorical, and the price of semiotics: an essay on philosophy of language and the doctrine of signs", *Semiotica* 161–1/4 (2006), 9–74.

2007. "The Primary Modeling System in Animals", in *La Filosofia del Linguaggio come arte dell'ascolto: sulla ricerca scientifica di Augusto Ponzio/Philosophy of Language as the art of listening: on Augusto Ponzio's scientific research*, ed. Susan Petrilli (Bari, Italy: Edizione dal Sud, 2007), 161–179.

2007a. *Intentionality and Semiotics. A tale of mutual fecundation* (Scranton, PA: Scranton University Press).

2007/8. "The Quo/Quod Fallacy in the Discussion of Realism", *Człowiek W Kulturze. Pismo poświęcone filozofii i kulturze*, Part 1 in vol. 19 (2007), 389–425; Part 2 in vol. 20 (2008), 289–316.

2008. *Descartes & Poinsot: the crossroad of signs and ideas*, Volume 2 of a "Postmodernity in Philosophy" Poinsot trilogy (Scranton, PA: University of Scranton Press).

2008a. "Evolution, semiosis, and ethics: rethinking the context of natural law", in *Contemporary Perspectives on Natural Law*, ed. Ana Marta González (Aldershot, England: Ashgate, 2008), 413–442. Reprinted in Cobley Ed. 2009: 74–90

2009. *Realism for the 21st Century. A John Deely Reader*, ed. Paul Cobley (Scranton, PA & London, UK: University of Scranton Press).

2009a. *Purely Objective Reality* (Berlin, Germany: Mouton de Gruyter).

2009b. *Augustine & Poinsot. The protosemiotic development*, Volume 1 of the "Postmodernity in Philosophy" Poinsot trilogy (Scranton, PA: University of Scranton Press).

2009c. "In The Twilight of Neothomism, a Call for a New Beginning. A return in philosophy to the idea of progress by deepening insight rather than by substitution", *American Catholic Philosophical Quarterly* 83.2 (Spring 2009), 267–278.

2010. "Realism and Epistemology", in *The Routledge Companion to Semiotics*, ed. Paul Cobley (London, UK: Routledge), 74–88.

2010a. *Semiotic Animal* (South Bend, IN: St Augustine's Press).

2010b. *Medieval Philosophy Redefined* (Scranton, PA & London, UK: University of Scranton Press).

2010c. "Projecting into Postmodernity Aquinas on Faith and Reason", *Addendum* to Chapter 8 of 2010b: 279–301.

2011. "Taking Faith Seriously", *Revue roumaine de philosophie*, tome 55, nr. 2.

2011? "Toward a Postmodern Recovery of 'Person'" (from the Proceedings of the 13–15 May International Congress "A Depersonalized Society? Educational Proposals" held at Universitat Abat Oliba CEU, Barcelona, Spain; forthcoming).

DEELY, John, Susan PETRILLI, and Augusto PONZIO.

2005. *The Semiotic Animal* (Ottawa, Canada: Legas Publishing).

DEELY, John N., and Raymond J. NOGAR.

1973. *The Problem of Evolution. A study of the philsophical repercussions of evolutionary science* (New York, NY: Appleton-Century-Crofts).

DEELY, John N., Brooke WILLIAMS, and Felicia E. KRUSE, Editors.

1986. *Frontiers in Semiotics* (Bloomington: Indiana University Press). Preface titled "Pars Pro Toto", pp. viii–xvii; "Description of Contributions", pp. xviii–xxii.

DENNETT, Daniel C.
  2006. *Breaking the Spell: Religion as a Natural Phenomenon* (New York: Viking Press).

DENZINGER.
  This name is used as an acronym for the *Enchiridion Symbolorum, definitionum et declarationum de rebus fidei et morum* (Compendium or 'Handbook' of the creeds and church doctrinal decisions), commissioned by Pope Pius IX, was compiled and published by Heinrich Denzinger (10 October 1819–1883 June 19) in 1854 (see Internet Archive <http://www.archive.org/details/enchiridionsymb 00creegoog>). Six editions appeared in Denzinger's lifetime. (Online biography at <http://www.catholic.org/encyclopedia/view.php?id=3769>.)
  After Denzinger's death in 1883 various editors continued to add to the work, and the numbering of documents changes from the original editions in the more recent editions (see the English *Wikipedia* entry "Heinrich Joseph Dominicus Denzinger" at <http://en.wikipedia.org/wiki/>, or the German entry "Enchiridion Symbolorum" at <http://de.wikipedia.org/wiki/Enchiridion_Symbolorum>). An English translation which includes documents to 1950 and retains the old numbering is available at <http://www.catecheticsonline.com/SourcesofDogma.php>
  The latest edition of the *Enchiridion* is the 42nd edition (ISBN 3-451-28520-7), by Peter Hünerman (Freiburg, Basel, Vienna: Herder, 2009).

DESCARTES, René.
  1641. *Meditations on First Philosophy*, trans. John Veitch in *The Rationalists* (Garden City, NY: Doubleday & Co. Anchor Books, 1974), 97–175.

DEUTSCH, David.
  1997. *The Fabric of Reality* (New York, NY: Penguin Books).

DeWITT, Bryce.
  2010. "Review of David Deutsch's *The Fabric of Reality* at <http://naturalscience.com/ns/books/book02.html>.

DRAKE, Stillman, Trans. and Editor.
    1957. *Discoveries and Opinions* (Garden City, NY: Doubleday). Translations of selections from Galileo's chief works, with introductions.

DUCROCQ, Albert.
    1957. *The Origins of Life*, trans. Alec Brown (London: Elek Books).

THE ECONOMIST.
    2010. "Ye cannae change the laws of physics — Or can you?" at <http://www.economist.com/node/16941123/print>.

EINSTEIN, Albert.
            Note: 1930, 1939, 1941, & 1948 below are available online at <http://www.sacred-texts.com/aor/einstein/einsci.htm>.
    1930. "Religion and Science", *New York Times Magazine* (9 November), 1–4; reprinted in Einstein 1954: 36–40.
    1934. "The Religious Spirit of Science", as reprinted in Einstein 1954: 40.
    1939. "Science and Religion, I", address of 19 May to Princeton Theological Seminary as reprinted in Einstein 1954: 41–44.
    1941. "Science and Religion, II", from a symposium on "Science, Philosophy and Religion" as reprinted in Einstein 1954: 44–49.
    1948. "Religion and Science: Irreconcilable?", *The Christian Register* (June 1948), as reprinted in Einstein 1954: 49–53.
    1951. "The Need for Ethical Culture", as reprinted in Einstein 1954: 53–54.
    1954. *Ideas and Opinions*, based on *Mein Weltbild*, ed. Carl Seelig (Zurich, Switzerland: Europa Verlag, 1953), together with other sources, with new translations and revisions by Sonja Bargmann (New York: Bonanza Books, 1954).

EMERY, O.P., Gilles.
    2007. *The Trinitarian Theology of Saint Thomas Aquinas*, trans. Francesca Murphy (New York: Oxford University Press).

FABRE, Jean-Henri.
1914. *The Mason Bees*, trans. by Alexander Teixeira de Mattos (New York: Garden City).

FABRO, Cornelio.
1968. *God in Exile: Modern Atheism; a study of the internal dynamic of modern atheism, from its roots in the Cartesian cogito to the present day*, trans. and ed. Arthur Gibson (Westminster, MD: Newman Press).

FANTOLI, Annibale (1924–).
1996. *Galileo. For Copernicanism and For the Church*, trans. George V. Coyne (2nd ed., rev. and corr.; Rome: Vatican Observatory Publications).
2003. *Galileo. For Copernicanism and For the Church*, trans. George V. Coyne (3rd ed., rev., corr., and expanded; Rome: Vatican Observatory Publications).

FAVARO, Antonio (1847–1922), Editor.
1890–1909. *Le opere di Galileo Galilei*, edizione nazionale sotto gli auspicii di Sua Maestà il re d'Italia (Firenze: Giunti Barbèra; ristampa 1929–1939), 20 Volumes. Includes important documents bearing on Galileo's life and work.

FESER, Edward.
2009. *Aquinas. A Beginner's Guide* (Oxford, England: One World Publications).

FEYNMAN, Richard P.,
1965 December 11. Nobel Lecture, "The Development of the Space-Time View of Quantum Electrodynamics"; online at <http://nobelprize.org/nobel_prizes/physics/laureates/1965/feynman-lecture.html>.
1985. *QED, The Strange Theory of Light and Matter* (Princeton, NJ: Princeton University Press).

FINILI, Antoninus, OP.
1948. "Natural Desire", Part 1, *Dominican Studies* I, 311–359.
1949. "Natural Desire", Part 2, *Dominican Studies* II, 1–15.
1952. "New Light on Natural Desire", *Dominican Studies* V, 159–84.

FOUCAULT, Michel.
1970.   *The Order of Things: An Archaeology of the Human Sciences* (New York, NY: Pantheon Books).
FREUD, Sigmund (6 May 1856–1939 September 23).
1912/13.   *Totem and Taboo: Resemblances between the Psychic Lives of Savages and Neurotics*, authorized trans. with an introduction by A. A. Brill (New York, NY: Prometheus Books; new ed. 2000).
1918.   *Totem and Taboo; Resemblances between the Psychic Lives of Savages and Neurotics*, authorized trans. by James Strachey (New York, NY: Moffat, Yard and Company) of 1913 German original.
1930.   *Civilization and Its Discontents*, authorized trans. by James Strachey (New York: Jonathan Cape & Harrison Smith) of 1929 German original.
1939.   *Moses and Monotheism*, trans. from the German by Katherine Jones (New York: Vintage Books, 1955).
GALILEO GALILEI (15 FEBRUARY 1564–1642 JANUARY 8).
Note: For the classic edition of Galileo's writings and related materials, see Favaro 1890–1909.
1615.   "Letter to the Grand Duchess Christina of Tuscany", trans. in Drake, ed. 1957: 173–216; original Italian in Favaro ed. 1968: vol. 5, pp. 309–348. Online see <http://www.disf.org/en/documentation/03-Galileo_Cristina.asp>
1623.   *Il Saggiatore* (Rome: Accademia dei Lincei), reprinted in *Galileo Galilei Opere*, con note de Pietro Pagnini Volume I (Florence, Italy: Casa Editrice A. Salani, 1964), pp. 103–437. Partial English trans. in Stillman Drake, Ed., *Discoveries and Opinions* (Garden City, NY: Doubleday, 1957), 231–280 (online at <http://www.princeton.edu/~hos/h291/assayer.htm>). A 1998–1999 partial trans. by George MacDonald Ross can be seen at <http://www.philosophy.leeds.ac.uk/GMR/hmp/texts/modern/galileo/assayer.html>.
GARDNER, Martin.
2004.   *Are Universes Thicker Than Blackberries?* (New York: W. W. Norton).

GARRIGAN, O. W.
 1967. "3. Theological Aspect" within the "Evolution, Human" 3-part article by E. L. Boné, R. J. Nogar, and O. W. Garrigan (pp. 676–685), in *New Catholic Encyclopedia* (1st ed.; New York: McGraw-Hill, 1967–1989), Vol. 5, pp. 684–685.

GAUTHIER, René Antoine.
 1970. "Introduction", being Tome I, vol. 1 ("Première Partie"), of *L'Éthique à Nicomaque*, traduction et commentaire par René Antoine Gauthier et Jean Yves Jolif (2nd ed. avec une Introduction nouvelle par Gauthier; Paris: Béatrice-Nauwelaerts), 2 tomes in 4 volumes (Introduction, Traduction, Commentaire livres i–v, Commentaire livres vi–x).

GENOVESI, SJ, Vincent J.
 1996. *In Pursuit of Love. Catholic Morality and Human Sexuality* (2nd ed.; Collegeville, MN: The Liturgical Press).

GENZ, Henning.
 1999. *Nothingness: the Science of Empty Space*, trans. Karin Heusch (Reading, MA: Perseus Books).

GILMORE, Michael R.
 1997. "Einstein's God. Just What Did Einstein Believe About God?", *Skeptic* vol. 5, no. 2, pp. 62ff.; online under "Einstein the Agnostic" <http://www.skeptically.org/thinker sonreligion/id8.html>.

GLATZ, Carol.
 2008. "God Made Pre-Humans into People, Vatican Newspaper Says", *Catholic News Service* 10 May 2008 at <http://www.catholicnews.com/data/stories/cns/0802496.htm>.

GOLDSTEIN, Rebecca.
 2006. *Betraying Spinoza; The Renegade Jew Who Gave Us Modernity* (New York: Schocken). Review in Bloom 2006.

GORDON, Jr., Raymond G., Editor.
 2005. *Ethnologue: Languages of the World* (15th ed.; Dallas, TX: SIL International; online version: <http://www.ethno logue.com/>.

GOULD, Stephen Jay.
　1997.　"The Evolution of Life on Earth", at <http://eddieting .com/eng/originoflife/gould.html> (under <http://edd ieting.com/eng/index.html>).
　2002.　*The Structure of Evolutionary Theory* (Cambridge, MA: Belknap Press).

GREDT, OSB, Jospehus.
　1924.　*De Cognitione Sensuum Externorum* (Rome: Desclée & Socii).
　1961.　*Elementa Philosophiae Aristotelico-Thomisticae* (13th ed. recognita et aucta ab E. Zenzen, OSB; Barcelona: Herder).

GROLEAU, Rick (Managing Editor of NOVA online).
　2003.　"Imagining Other Dimensions", in *Nova: The Elegant Universe*, at <http://www.pbs.org/wgbh/nova/elegant/ dimensions.html>.

GUAGLIARDO, OP, Vincent.
　2011.　"Father and Son in the Trinity: Metaphor or Analogy?", *The American Journal of Semiotics* 27.1–4, in press.

GUTH, Alan H.
　1981.　"Inflationary Universe: A possible solution to the ho- rizon and flatness problems", *Physical Review D* 23.2 (1981), 347–356. (At http://prd.aps.org/abstract/PRD/ v23/i2/p347_1)
　1997.　*The Inflationary Universe: The Quest for a New Theory of Cosmic Origins* (Reading, MA: Addison-Wesley).

HANNAM, James.
　2009.　*God's Philosophers: How the Medieval World Laid the Foun- dations of Modern Science* (London, UK: Icon Books).

HARDWICK, Charles S. Editor, with the assistance of James Cook.
　1977.　*Semiotics and Significs. The Correspondence between Charles S. Peirce and Victoria Lady Welby* (Bloomington: Indiana University Press).

HARRIS, Sam.
　2004.　*The End of Faith: Religion, Terror, and the Future of Reason* (New York: W. W. Norton & Co.).

HART, Stephen.
　1996.　*The Language of Animals* (New York: Henry Holt & Co.).

2010. "The Animal Communication Project", at <http://acp.eugraph.com/monkey/index.html>.

HAWKING, Stephen (8 January 1942–), and Leonard MLODINOW (1954–).
2010. *The Grand Design* (New York, NY: Bantam Books).

HEGEL, G. W. F.
1830. *Enzyklopädie der philosophischen Wissenschaften*, neu herausgegeben von Friedhelm Nicolin und Otto Pöggeler (Hamburg: Felix Meiner, 1959).

HEINRICH, Bernd.
1999. *The Mind of the Raven: Investigations and Adventures with Wolf-Birds* (New York, NY: HarperCollins e-books).

HENDERSON, L. J.
1958. *The Fitness of the Environment* (Boston: Beacon).

HERBERT, Nick.
1987. *Quantum Reality: Beyond the New Physics* (Garden City, NY: Anchor/Doubleday).

HETTCHE, Matt.
2008. "Christian Wolff", *The Stanford Encyclopedia of Philosophy* (Fall 2008 Edition), ed. Edward N. Zalta, URL = <http://plato.stanford.edu/archives/fall2008/entries/wolff-christian/>.

HIMMA, Kenneth Einar.
2009. "Design Arguments for the Existence of God", online at <http://www.iep.utm.edu/design/>.

HITCHENS, Christopher.
2007. *God Is Not Great. How religion poisons everything* (Boston, MA: Twelve Books, of the Hachette Book Group).

HOLTON, Gerald.
2002. "Einstein's Third Paradise", Daedelus (Fall 2002), 26–34; online at <http://www.aip.org/history/einstein/essay-einsteins-third-paradise.htm>.

HORGAN, John.
1996. *The End of Science* (New York: Broadway reprint).

HOROWITZ, Norman H.
   1956.   "The Origin of Life", in *Engineering and Science* 20.2, pp.
          21–25.

HOWARD, Don.
   2003.   "Two Left Turns Make a Right: On the Curious Politi-
          cal Career of North American Philosophy of Science at
          Mid-century", in *Logical Empiricism in North America*,
          ed. Alan Richardson and Gary Hardcastle (Minneapolis,
          MN: University of Minnesota Press), 25–93.

HUANG, S.
   1959.   "Occurrence of Life In the Universe", *American Scientist*
          47 (September, 1959), 397–402.

INTERNET ENCYCLOPEDIA OF PHILOSOPHY.
   2010a. "Rudolph Carnap" at <http://www.iep.utm.edu/c/carnap
          .htm>.
   2010b. "Vienna Circle" at <http://www.iep.utm.edu/v/viennaci
          .htm>.

IRVINE, A. D.
   2010.   "Bertrand Russell", *The Stanford Encyclopedia of Philoso-
          phy* (Winter 2010 Edition), Edward N. Zalta, ed., URL
          = <http://plato.stanford.edu/archives/win2010/entries/
          russell/>.
   2009.   "Russell's Paradox", *The Stanford Encyclopedia of Philoso-
          phy* (Summer 2009 Edition), Edward N. Zalta, ed., URL
          = <http://plato.stanford.edu/archives/sum2009/entries/
          russell-paradox/>.

JAKI, Stanley.
   1985.   *God and the Cosmologists* (Edinburgh: Scottish Academic
          Press).
   2010.   Jaki bibliography <http://theduhemsociety.blogspot.com
          /2009/04/bibliography-of-stanley-l-jaki.html>.

JAMMER, Max.
   1999.   *Einstein and Religion* (Princeton, NJ: Princeton Univ.
          Press, 1999).

JOHNSON, George.
   2006.   "A Free-for-All on Science and Religion", *New York
          Times, Science Times* (11/21/2006), reporting on a Con-

ference held at the Salk Institute for Biological Studies, La Jolla, CA.

JOURNET, Charles, Cardinal.
1955. *The Church of the Word Incarnate: an Essay in Speculative Theology* (London: Sheed & Ward), 3 Vols., being the English trans. by A. H. C. Downes of *L'Eglise du verbe incarné: Essai de théologie spéculative* (Paris, France: Desclée, 1941), 4 Vols.

KAUFFMAN, Stuart A.
1995. *At Home in the Universe: The Search for the Laws of Self-Organization and Complexity* (New York: Oxford University Press).

KOSSEFF, Lauren.
2010. "Primate Use of Language", at <http://www.pigeon.psy .tufts.edu/psych26/language.htm>.

KRETZMANN, Norman.
1997. *The Metaphysics of Theism: Aquinas's Natural Theology in Summa Contra Gentiles I* (Oxford, UK: Clarendon Press).
1999. *The Metaphysics of Creation: Aquinas's Natural Theology in Summa Contra Gentiles II* (Oxford, UK: Clarendon Press).

KIRSTIĆ, K.
1964. "Marko Marulic [1450–1524] — The Author of the Term 'Psychology'," *Acta Instituti Psychologici Universitatis Zagrabiensis*, no. 36 (1964), pp. 7–13. See online <http://psychclassics.yorku.ca/Krstic/marulic.htm>

KÜNG, Hans, Josef VAN ESS, Heinrich STIETENCRON, and Heinz BECHERT.
1993. *Christianity and World Religions: Paths to Dialogue with Islam, Hinduism, and Buddhism* (Maryknoll, New York: Orbis Books).

KURZWEIL, Ray.
2001. "The Law of Accelerating Returns", posted 7 March 2001 on *Kurzweil Accelerating Intelligence* at <http://www .kurzweilai.net/the-law-of-accelerating-returns>.

LAURENTIN, René.
    1991.   *A Short Treatise on the Virgin Mary*, trans. Charles Neu-
             mann, S.M. (Washington, NJ: AMI Press), pp. 328–329.

LEMONICK, Michael D.
    2001.   "How the Universe Will End", *Time* magazine (cover
             story; 25 June 2001); online at <www.time.com/time/cov-
             ers/1101010625/story.html> or <http://www.time.com
             /time/magazine/article/0,9171,1000170,00.html>.
    2004.   "Before the Big Bang", *Discover* (February 2004), 35–41.
             Published online 5 February 2004 <http://discoverma
             gazine.com/2004/feb/cover>; available as downloadable
             PDF at <http://www.physics.princeton.edu/~steinh/
             Discover0204.pdf>.

LIENHARD, John H.
    1989.   "Inventing Benzene", Episode 265 of *Engines of Our In-
             genuity* at <http://www.uh.edu/engines/epi265.htm>.

LINDBERG, David C., and Ronald L. NUMBERS, Editors.
    2003.   *When Science and Christianity Meet* (Chicago; University
             of Chicago Press, 2003).

LINDE, A. D.
    1990.   *Particle Physics and Inflationary Cosmology* (Chur, Switzer-
             land: Harwood Academic Publishers).

LINDER, Doug.
    2004.   "The Vatican's View of Evolution: The Story of Two
             Popes", online at <http://www.law.umkc.edu/faculty/proj-
             ects/ftrials/conlaw/vaticanview.html>, part of "The Evo-
             lution, Creationism, and Intelligent Design Controversy"
             from the "Evolution/Creationism Homepage" of "Explor-
             ing Constitutional Conflicts" at <http://www.law.umkc
             .edu/faculty/projects/ftrials/conlaw/evolution.html>.

LINDLEY, David.
    1993.   *The End of Physics: The Myth of a Unified Theory* (New
             York, NY: Basic Books).

LUCAS, John Randolph.
    1961.   "Minds, Machines and Gödel", *Philosophy* XXXVI, pp.
             112–127. Reprinted in *The Modeling of Mind*, ed. Kenneth
             M. Sayre and Frederick J. Crosson (South Bend, IN:

University of Notre Dame Press, 1963), 269–270; and in *Minds and Machines*, ed. Alan Ross Anderson (New York, NY: Prentice-Hall, 1954), 43–59.

1970. *The Freedom of the Will* (Oxford University Press).

MAHMOOD, Nawal.
2011. "Laws of Nature May Not Be the Same Everywhere. Revolution in Physics?". *TheTechJournal* (Technological News Portal, Saturday 12 February 2011), online at <http://thetechjournal.com/science/laws-of-nature-may-not-be-the-same-everywhere-revolution-in-physics.xhtml>.

MARITAIN, Jacques.
1943. "The Natural Mystical Experience and the Void", in his *Redeeming the Time* (London: Geoffrey Bles: The Centenary Press), 225–255.

1959. *Distinguish to Unite, or The Degrees of Knowledge*, trans. from the 4th French ed. of *Distinguer pour Unir: Ou, les Degrés du Savoir* (Paris: Desclée de Brouwer, 1932) under the supervision of Gerald B. Phelan (New York: Scribner's).

1967. "Toward a Thomist Idea of Evolution", *Nova et Vetera* 2, 87–136.

1977. *On the Church of Christ* (Notre Dame, IN: University of Notre Dame Press, 1977), trans. by Joseph W. Evans of *De l'Église du Christ* (Paris: Desclée de Brouwer, 1970).

MARTIN, Christopher.
1997. *Thomas Aquinas. God and Explanations* (Edinburgh, Scotland: Edinburgh University Press).

MARTIN, Michael, and Ricki MONNIER, Eds.
2003. *The Impossibility of God* (New York: Prometheus).

MERKLE, Ralph C.
1988. "How Many Bytes in Human Memory", *Foresight Update* No. 4 (October); online at <http://www.merkle.com/humanMemory.html>.

1989. "Energy Limits to the Computational Power of the Human Brain", *Foresight Update* No. 6 (August); online at <http://www.merkle.com/brainLimits.html>.

MILLER, F.S.E., Sister Paula Jean.
  1996. *Marriage: The Sacrament of Divine-Human Communion: A Commentary on St. Bonaventure's Breviloquium* (Quincy, IL: Franciscan Press).

MOST, William G. (13 August 1914–1999 January 21).
  2010. "Our Lady's Physical Virginity in the Birth of Jesus", at <http://www.ewtn.com/library/SCRIPTUR/virbir.htm>.

MULHOLLAND, OFM, Seamus.
  2001. "Incarnation in Franciscan Spirituality", at <http://www.franciscans.org.uk/2001jan-mulholland.html> or <http://franciscans.beimler.org/Incarnation%20Spirit ality.html>.

MURDOCK, George P., and Douglas R. WHITE.
  1969. "Standard Cross-Cultural Sample: on-line", *Ethnology* 8, pp. 329–369. Cf. <http://worldcultures.org/>.
  2008. "Standard Cross-Cultural Sample: on-line" of 1969 reprinted with Addenda and Footnotes, available online in downloadable PDF format at <http://eclectic.ss.uci .edu/~drwhite/pub/SCCS1969.pdf>; Appendix A, the list of societies in Table 1 and data on these societies, at <http://eclectic.ss.uci.edu/~drwhite/worldcul/Sccs34 .htm>; list of variables for the societise coded at <http:// eclectic.ss.uci.edu/~drwhite/SCCS/>; and societies on the map for the standard sample at <http://eclectic.ss .uci.edu/~drwhite/worldcul/sccs.html>.

MURZI, Mario.
  2001. "Rudolf Carnap (1891–1970)", *The Internet Encyclopedia of Philosophy* <http://www.iep.utm.edu/carnap/>.
  2004. "Vienna Circle", *The Internet Encyclopedia of Philosophy* <http://www.iep.utm.edu/viennacr/>.

NASA (NATIONAL AERONAUTICS AND SPACE ADMINISTRA- TION) WEBSITE TAB "UNIVERSE", 10 December 2010 update.
  2010. "What is the Universe Made Of?" <http://map.gsfc.nasa .gov/m_uni/uni_101matter.html>.

NEWTON, Isaac.
  1692. Quoting his letter to Richard Bentley dated 12 October 1692 in his "Preface" to Bentley's 1773 *Observations on the Prophecies of Daniel and the Apocalypse of St. John*, (London: Darby & Browne, 1733), reprinted as *The Prophecies of Daniel and the Apocalypse* (Hyderabad, India: Printland Publishers, 1998).

NOGAR, Raymond J.
  1963. *The Wisdom of Evolution* (New York, NY: Doubleday).
  1966. *The Lord of the Absurd* (New York, NY: Herder and Herder).
  1967. "2. Philosophical Aspect" within the "Evolution, Human" 3-part article by E. L. Boné, R. J. Nogar, and O. W. Garrigan (pp. 676–685), in *New Catholic Encyclopedia* (1st ed.; New York: McGraw-Hill, 1967–1989), Vol. 5, pp. 682–684.
  1967a. "Evolution, Organic", *loc. cit.* pp. 685–694.
  1967b. "Evolutionism", *loc. cit.* pp. 694–696.

O'BRIEN, Thomas C.
  1960. *Metaphysics and the Existence of God* (Washington, DC: The Thomist Press).

O'MEARA, Thomas F.
  1997. *Thomas Aquinas, Theologian* (Notre Dame, IN: University of Notre Dame Press).

OMNÈS, Roland.
  1999. *Quantum philosophy: understanding and interpreting contemporary physics*, trans. Arturo Sangalli (Princeton, N.J.: Princeton University Press). (See <http://libcat.slu.edu /search/aOmn%7bu00E8%7ds%2C+Roland./aomnes +roland/-2,-1,0,B/browse>.)
  2005. *Converging Realities : toward a Common Philosophy of Physics and Mathematics* (Princeton, N.J.: Princeton University Press).

OPARIN, A. I.
  1957. *The Origin of Life on Earth* (3rd ed.; New York: Macmillan).

PARRY-HILL, Matthew J., and Michael W. DAVIDSON.
2006.   "Thomas Young's Double Slit Experiment", an interactive Java tutorial online with Molecular Expressions™ at <http://micro.magnet.fsu.edu/primer/java/interference/doubleslit/>.

PARSONS, Talcott, and Edward SHILS, Editors.
1951.   *Toward a General Theory of Action* (Cambridge, MA: Harvard University Press).

PATTERSON, Francine, and Eugene LINDEN.
1985.   *The Education of Koko* (New York, NY: Holt, Rinehart and Winston); see esp. Chapter 3, "Gorrilla Gorilla", online at <http://www.koko.org/world/teok_ch3.html>. Also "Talking with Chimps" at <http://www.geocities.com/RainForest/Vines/4451/TalkWithChimps/>.

PEIRCE, Charles Sanders.
i.1866–1913.   *The Collected Papers of Charles Sanders Peirce* (abbreviated "CP"), in 8 vols., ed. C. Hartshorne, P. Weiss, and A. W. Burks (Cambridge, MA: Harvard University Press, 1931–1958); all eight vols. in electronic form ed. John Deely with an Introduction "Membra Ficte Disjecta — A Disordered Array of Severed Parts" (Charlottesville, VA: Intelex Corporation, 1994). Dating within the CP volumes is based on the Burks Bibliography at the end of volume 8. The abbreviation followed by volume and paragraph numbers with a period between follows the standard CP reference form.

i.1867–1913.   *The Essential Peirce (1867–1893), Volume 1* (abbreviated "EP"), ed. Nathan Houser and Christian Kloesel (Bloomington, IN: Indiana University Press, 1992); *The Essential Peirce (1893–1913), Volume 2*, ed. Nathan Houser, André De Tienne, Jonathan R. Eller, Cathy L. Clark, Albert C. Lewis, D. Bront Davis (Bloomington, IN: Indiana University Press, 1998). Volume 1 duplicate materials from CP, but volume 2 consists of previously unpublished manuscripts. The abbreviation EP followed by volume and page number(s) is a standardized reference form.

c.1902. "Minute Logic", draft for a book complete consecutively only to Chapter 4. Published in CP in extracts scattered over six of the eight volumes, including 1.203–283, 1.575–584; 2.1–202; 4.227–323, 6.349–352; 7.279, 7.374n10, 7.362–387 except 381n19. (For fuller detail, see Burks 293–294.)

c.1905. "Pragmaticism, Prag. [4]". First page of ms. is missing; printed in CP 5.502–537.

1908. Draft of a letter to Lady Welby dated December 24, 25, 28 "On the Classification of Signs", CP 8.342–379 except 368n23 are from it (Burks p. 321 par. 20.b). In Hardwick ed. 1977: 73–86; and EP 2.478–483.

PERIODIC TABLE.

2010. "WebElements™ periodic table" at <http://www.webe lements.com/>.

PETRILLI, Susan, and Augusto PONZIO.

2003. *Semioetica* (Rome: Meltemi).

PINKER, Stephen.

1994. *The Language Instinct: How the Mind Creates Language* (New York, NY: HarperCollins). See comment at <http:// en.wikipedia.org/wiki/The_Language_Instinct>.

PITTENDRIGH, Colin S.

1958. "Adaptation, Natural Selection, and Behavior", in *Behavior and Evolution*, ed. Anne Roe and George Gaylord Simpson (New Haven, CT: Yale University Press), pp. 390–416.

POINSOT, John (*Joannes a Sancto Thoma*, "John of St. Thomas").

1632. *Tractatus de Signis*, bilingual ed. by John Deely with Ralph A. Powell (1st independent ed.; Berkeley: University of California Press, 1985).

1637–1667. *Cursus Theologicus*, in 8 volumes (vol. 1 Alcalá, Spain: 1637; vols. 2 & 3 Lyons, France: 1643; vol. 4–7, ed. Didacus Ramirez, Madrid, Spain — 4 & 5 in 1645, 6 in 1649, 7 in 1656; vol. 8, ed. Franciscus Combefis, Paris). The Vivès edition reprints the original 8 volumes as 9 volumes, Paris: 1885. A critical edition by Dom Edmond Boissard, O.S.B., of Solesmes Abbey (1 January

1898–1979 December 11), in 5 volumes, gets only as far as original volume 4.

1643.   "Tractatus de Angelis" in *Joannis a Sancto Thoma Cursus Theologicus Tomus IV*, Solesmes ed. (Paris: Desclée, 1946), pp. 441–835.

PORPHYRY THE PHOENICIAN.

c.AD271.   *Porphyrii Isagoge et in Aristotelis Categorias Commentarium* (Greek text), ed. A. Busse (Berlin, 1887); English trans. by Edward W. Warren, *Porphyry the Phoenician: Isagoge* (Toronto: Pontifical Institute of Mediaeval Studies, 1975).

POWELL, Ralph Austin (21 September 1914–2001 June 12).

1982.   *Freely Chosen Reality* (Washington, D.C.: University Press of America).

PRICE, Michael Clive.

1995.   "Many Worlds Quantum Theory", at <http://kuoi.asui .uidaho.edu/~kamikaze/doc/many-worlds-faq.html>.

RAHNER, S.J., Karl (20 March 1904–1984 March 30).

1978.   *Foundations of Christian Faith. An introduction to the idea of Christianity* (New York, NY: Seabury Press), trans. by William V. Dych of *Grundkurs des Glaubens: Einführung in den Begriff des Christentums* (Freiburg im Breisgau, Germany: Verlag Herder, 1976).

1984.   "The Experiences of a Catholic Theologian", *Communio* 11.4 (1984), 404–414.

RATZINGER, Joseph (16 March 1927– ; Pope Benedict XVI as of 19 April 2005).

1988.   *Eschatology, Death and Eternal Life*, trans. Michael Waldstein, ed. Aidan Nichols, OP (Washington, DC: Catholic University of America Press).

2007.   2nd printing in English of 1988, with an added a "Preface" by Peter Casarella, pp. xi–xiii, and a new "Foreword" xvii–xxii by the author as now Pope Benedict XVI.

REES, Sir Martin.

2004.   *Our Final Century: Will the Human Race Survive the Twenty-first Century?* (New York, NY: Basic Books).

REUCROFT, Stephen, and John SWAIN.
   1998. "What is the Casimir Effect?", *Scientific American* (22 June 1998), p. 8.

RIZZI, Anthony.
   2004. *The Science Before Science; A Guide to Thinking in the 21st Century* (Bloomington, IN: AuthorHouse). Cf. <http://www.iapweb.org/>.
   2008. *Physics for Realists: Mechanics — Modern physics with a common sense grounding* (Baton Rouge, LA: Press of the Institute for Advanced Physics).

ROCKMORE, Tom, and Beth J. SINGER, Editors.
   1992. *Antifoundationalism Old and New* (Philadelphia, PA" Temple University Press).

RUSSELL, Bertrand.
   1918. "The Study of Mathematics", Chap. 4 in *Mysticism and Logic and Other Essays* (London: Longmans, Green), pp. 60–73.

SAGAN, Carl.
   1980. *Cosmos* series: <http://www.hulu.com/cosmos>.
   1996. *The Demon-Haunted World: Science as a Candle in the Dark* (New York: Ballantine Books). See <http://en.wikipedia.org/wiki/The_Demon-Haunted_World>.

SATO, Rebecca.
   2009. "Is Dark Matter & Dark Energy the Same Thing?", *The Daily Galaxy* (5 February 2009), at <http://www.dailygalaxy.com/my_weblog/2009/02/could-dark-matt.html>.

SCHOONENBERG, S.J., Piet.
   1965. *Man and Sin: A Theological View*, trans. Joseph Donceel (South Bend, IN: University of Notre Dame Press).
   1967. "Original sin and man's situation", *Theology Digest* 15.3 (Autumn), 203–208.

SCIENCE DAILY.
   2009. "Surprising Changes in Black Hole-powered 'Blazar' Galaxy"; Retrieved January 30, 2011, from <http://www.sciencedaily.com-/releases/2009/03/090318162624.htm>.

SCIENCE AND THEOLOGY NEWS, On-line Edition, "Many Worlds", http://www.stnews.org.

SCHMITZ, Kenneth L.

2010.   "Semiotics or Metaphysics as First Philosophy?", *Semiotica* 179–1/4, 119–132. This essay is in the 2nd of the 2-volume Semiotica Special Issue Guest-Edited by Susan Petrilli and John Hittinger discussing Deely's 2001 *Four Ages of Understanding*.

SEBEOK, Thomas A.

1963.   "Book review article of M. Lindauer, *Communication among Social Bees*; W. N. Kellog, *Porpoises and Sonar*; and J. C. Lilly, *Man and Dolphin*", *Language* 39, 448–466.

1968.   "Goals and Limitations of the Study of Animal Communication", in Sebeok ed. 1968: 3-14, q.v.; reprinted in Sebeok 1985: 59–69, to which page reference is made.

1975a.   "Zoosemiotics: At the Intersection of Nature and Culture", in *The Tell-Tale Sign*, ed. T. A. Sebeok (Lisse, the Netherlands: Peter de Ridder Press), pp. 85–95.

1975b.   "Six Species of Signs: Some Propositions and Strictures", *Semiotica* 13.3, 233–260, reprinted in Sebeok 1985: 117–142.

1977.   "Zoosemiotic Components of Human Communication", in *How Animals Communicate*, ed. Thomas A. Sebeok (Bloomington: Indiana University Press), Chap. 38, pp. 1055–1077.

1978.   "'Talking' with Animals: Zoosemiotics Explained", *Animals* 111.6 (December), pp. 20ff.

1978a.   "Looking in the Destination for What Should Have Been Sought in the Source", *Diogenes* 104, 112–138; reprinted in Sebeok 1989: 272–279.

1984, June 3.   "The Evolution of Communication and the Origin of Language", lecture in the June 1–3 ISISSS '84 Colloquium on "Phylogeny and Ontogeny of Communication Systems". Published under the title "Communication, Language, and Speech. Evolutionary Considerations", in Sebeok 1986: 10–16.

2001.   *Global Semiotics* (Bloomington, IN: Indiana University Press).

SEBEOK, Thomas A., Editor.
  1968. *Animal Communication: Techniques of Study and Results of Research* (Bloomington, IN: Indiana University Press).
  1977. *How Animals Communicate* (Bloomington: Indiana University Press).

SEBEOK, Thomas A., General Editor; Paul BOUISSAC, Umberto ECO, Jerzy PELC, Roland POSNER, Alain REY, Ann SHUKMAN, Editorial Board.
  1986. *Encyclopedic Dictionary of Semiotics* (Berlin: Mouton de Gruyter), in 3 Volumes.

SEBEOK, Thomas A., and Robert ROSENTHAL, Editors.
  1981. *The Clever Hans Phenomenon: Communication with Horses, Whales, Apes, and People* (New York: The New York Academy of Sciences).

SEBEOK, Thomas A., and Jean UMIKER-SEBEOK, Editors.
  1980. *Speaking of Apes: A Critical Anthology of Two-Way Communication With Man* (New York: Plenum).

SEIFE, Charles.
  2005. "What Is the Universe Made Of?", *Science* 309.5731 (1 July 2005), p. 78; online at <http://www.sciencemag.org/content/309/5731/78.1.full?sid=641bb83b-40fe-44a2-8b71-4ce46f256f3a>.

SERTILLANGES, OP, Antonin-Dalmace.
  1945. *L'Idée de Creation* (Paris: Aubier).

SETI (= SEARCH FOR EXTRA-TERRESTRIAL INTELLIGENCE), at <http://www.space.com/searchforlife/>.

SHERWIN, O.P., Michael.
  1996. "Reconciling Old Lovers: John Paul on Science and Faith", *Catholic Dossier* 2.4 (July-August), 20–25. Online at <http://www.jp2forum.org/articles_about_pope/Culture_and_Education/Sherwin_on_Science_and_Faith.pdf>.

SINGER, Peter.
  1975. *Animal Liberation: A New Ethics for our Treatment of Animals* (New York, NY: Random House).

2006.   "The Ethics of Life" and "The Great Ape Debate", at <http://www.project-syndicate.org/series/31/description>.

SMART, Ninian.
1999.   *World Philosophies* (London: Routledge).

SMITH, Barry D.
2006.   "The Unmoved Mover in Physics" and "The Unmoved Mover in Metaphysics", online at <http://www.abu.nb.ca/Courses/GrPhil/PhilRel/Aristotle.htm>. See also <http://www.abu.nb.ca/courses/grphil/IndexGrPh.htm> and particularly <http://www.abu.nb.ca/courses/grphil/Aristotle.htm>.

SNOW, C. P.
1959.   *The Two Cultures and the Scientific Revolution* (New York: Cambridge University Press).

STACE, Walter Terence.
1955.   *The Philosophy of Hegel* (New York: Dover, 1955).

STENGER, Victor J.
2006.   *God: The Failed Hypothesis* (Amherst, NY: Prometheus Books).

STIMSON, Dorothy.
1972.   *The Gradual Acceptance of the Copernican Theory of the Universe* (Gloucester, MA: Peter Smith).

TCKACZ, Dr. Catherine Brown.
2002.   "Reproductive Science and the Incarnation", *Fellowship of Catholic Scholars Quarterly* Vol. 25, No. 4 (Fall 2002), 11–25.

TEGMARK, Max.
2007.   *Parallel Universes* at <http://www.relativitycalculator.com/articles/max_tegmark/parallel_universes_max_tegmark.html>. (Earlier pdf file at <http://space.mit.edu/home/tegmark/multiverse.pdf>.)

TEILHARD DE CHARDIN, Pierre.
1959.   *The Phenomenon of Man*, with an Introduction by Julian Huxley, trans. Bernard Wall (New York: Harper & Brothers), from *Le Phénomène Humain* (Paris: Editions du Seuil, 1955).

THEOPEDIA.
   2009.   "God of the gaps", in *Theopedia, an encyclopedia of Biblical Christianity* <http://www.theopedia.com/God_of_the_Gaps>.

TORRELL, Jean-Pierre, OP.
   1996.   *St. Thomas Aquinas*, vol. 1, *The Person and His Work*, trans. Robert Royal (Washington, DC: Catholic University of America Press).

TRAKAKIS, Nick.
   2004/5. "What Was the Iconoclast Controversy About?", *Theandros. An Online Journal of Orthodox Christian Theology and Philosophy*, vol. 2, no. 2 (Winter 2004/2005), at <www.theandros.com/iconoclast.html>.

TURNBULL, Robert G.
   1998.   *The* Parmenides *and Plato's Late Philosophy. Translation of and commentary on the* Parmenides *with Interpretative Chapters on the* Timaeus, *the* Theaetetus, *the* Sophist, *and the* Philebus (Toronto: Univeristy of Toronto Press).

TYSON, Peter.
   2010.   "Galileo's Battle for the Heavens; His Big Mistake", at <http://www.pbs.org/wgbh/nova/galileo/mistake.html>.

VITZ, Paul.
   1988.   *Sigmund Freud's Christian Unconscious* (New York: Guilford Press).

VON BAEYER, Hans Christian.
   1989.   "A Dream Come True", *The Sciences* 29.1 (January-February 1989), pp. 6–8.

WADDINGTON, Conrad Hal.
   1960.   "Evolutionary Adaptation", in *The Evolution of Life*, Vol. 1 of *Evolution after Darwin*, ed. Sol Tax (Chicago, IL: University of Chicago Press), pp. 381–402.
   1961.   "The Biological Evolutionary System", Chapter 9 of his *The Ethical Animal* (New York, NY: Atheneum), 84–100.

WALDMAN, Lev.
   2002.   "Many-Worlds Interpretation of Quantum Mechanics", *Stanford Encyclopedia of Philosophy* at <http://plato.stanford.edu/entries/qm-manyworlds/>.

WALLACE, O.P., William A.
1963.   *Einstein, Galileo, and Aquinas: Three Views of Scientific Method* (Washington, DC: Thomist Press).
WARD, Mark.
1997.   "End of the road for brain evolution", *New Scientist* vol. 153, no. 2066 (25 January 1997), p. 14. Online at <http://www.bio.net/bionet/mm/neur-sci/1997-January/027379.html>.
WEINBERG, Stephen.
1992.   *Dreams of a Final Theory* (New York: Pantheon Books, 1992)
1998.   "The Revolution that Didn't Happen", *The New York Review of Books*, Vol. XLV, No. 15 (October 8), pp. 48–52; online at <http://www.nybooks.com/articles/archives/1998/oct/08/the-revolution-that-didnt-happen/>.
1999.   "A Designer Universe", article is based on a talk given in April 1999 at the Conference on Cosmic Design of the American Association for the Advancement of Science in Washington, DC; online at <http://www.physlink.com/education/essay_weinberg.cfm>.
2002.   "Is the Universe a Computer?", review of Stephen Wolfram's *A New Kind of Science* (Champaign, IL: Wolfram Media). See <http://service.bfast.com/bfast/click7.bfmid=2181&sourceid=41397204&bfpid=1579550088>.
2005.   "Living in the Multiverse" at <http://arxiv.org/abs/hep-th/0511037v1>; subsequently published in Bernard Carr, ed., *Universe or Multiverse* (Cambridge, England: Cambridge University Press, 2007), pp. 29–42.
2010.   Untitled interview on God, at <http://www.pbs.org/faithandreason/transcript/wein-frame.html>.
WEISBERG, Robert.
1992.   "Friedrich August von Kekule", in *Creativity, Beyond the Myth of Genius* (New York: W. H. Freeman); see online <http://members.optusnet.com.au/~charles57/Creative/Brain/kekule.htm>.
2006.   *Creativity, Beyond the Myth of Genius* (New York; W.H. Freeman & Company).

WEISHEIPL, James A., O P.
  1974. *Friar Thomas d'Aquino: His Life, Thought and Work* (New York: Doubleday).

WELLS, Jonathan.
  1998. "Abusing Theology: Howard Van Till's Forgotten Doctrine of Creation's Functional Integrity", online as Discovery Institute Debate *Origins & Design* 19:1 at <http://www.arn.org/docs/odesign/od191/abusingtheology191.htm>.

WHITE, Richard.
  2007. Website "Abiogenesis — Origins of Life Research" under <http://darwiniana.org/abiogenesis.htm>.

WIKIPEDIA.
  2010a. "Accelerating Change" <http://en.wikipedia.org/wiki/Law_of_Accelerating_Returns>.
  2010b. "Aevum" <http://en.wikipedia.org/wiki/Æviternity>.
  2010c. "Brainstorming" <http://en.wikipedia.org/wiki/Brain-storming>.
  2010d. "Charles; Darwin's Illness" <http://en.wikipedia.org/wiki>.
  2010e. "Many World's Interpretation" <http://en.wikipedia.org/wiki/Many-worlds_interpretation>.
  2010f. "Origin of Life" <http://en.wikipedia.org/wiki/Origin_of_life>.
  2010g. "Philosophy of Mind" <http://en.wikipedia.org/wiki/Philosophy_of_mind>.
  2010h. "String Theory" <http://en.wikipedia.org/wiki/String_theory>.
  2011a. "Abiogenesis" <http://en.wikipedia.org/wiki/Abiogenesis>
  2011b. "Accelerating Change" <http://en.wikipedia.org/wiki/Accelerating_change>.
  2011c. "Animal Communication" <http://en.wikipedia/org/wiki/Animal-communication>.
  2011d. "Animal Language" <http://en.wikipedia.org/wiki/Animal_language>.
  2011e. "Brainstorming" <http://en.wikipedia.org/wiki/Brain storming>.
  2011f. "Carl Sagan" <http://en.wikipedia.org/wiki/Carl_Sagan>

2011g. "Charles Darwin's Health" <http://en.wikipedia.org/wiki/Charles_Darwin's_health>.

2011h. "Dark Energy" <http://en.wikipedia.org/wiki/Dark_energy>.

2011i. "Dark Matter" <http://en.wikipedia.org/wiki/Dark_matter>.

2011j. "God of the Gaps" <http://en.wikipedia.org/wiki/God_of_the_gaps>.

2011k. "Intelligent Design" <http://en.wikipedia.org/wiki/Intelligent_design>.

2011l. "The Language Instinct" <http://en.wikipedia.org/wiki/The_Language_Instinct>.

2011m. "Length Contraction" <http://en.wikipedia.org/wiki/Length_contraction>.

2011n. "Mass-Energy Equivalence" <http://en.wikipedia.org/wiki/E%3Dmc%C2%B2>.

2011o. "Miller-Urey Experiment" <http://en.wikipedia.org/wiki/Miller%E2%80%93Urey_experiment>.

2011p. "Negentropy" <http://en.wikipedia.org/wiki/Negentropy>.

2011q. "Observable Universe" <http://en.wikipedia.org/wiki/Observable_universe>, esp. the section "Misconceptions" <http://en.wikipedia.org/wiki/Observable_universe#Misconceptions>.

2011r. "String Theory" <http://en.wikipedia.org/wiki/String_theory>.

2011s. "Thomas A. Sebeok" <http://en.wikipedia.org/wiki/Thomas_A._Sebeok>.

2011t. "Transhumanism" <http://en.wikipedia.org/wiki/Transhumanism>.

WINTERS, Mark.

1993–2010. "The Periodic Table on the Word-Wide Web" <http://www.webelements.com/>.

WOIT, Peter.

2002. "Is String Theory even Wrong?", article in the "Macroscope" Department of *American Scientist* 90.2 (March-April 2002), p. 110; online at <http://www.americanscientist.org/issues/pub/2002/3/is-string-theory-even-wrong>.

ZYGON, Journal of Religion & Science.
    2011?   Publication of the Institute on Religion in an Age of Science (IRAS): see <http://www.blackwellpublishing.com/journal.asp?ref=0591-2385>.

INDEX

Colophon

Typesetting by Marty Klaif

Graphic design consultation with Czeslaw Jan Grycz

Composed in Adobe InDesign CS4 using Janson Text LT Pro

On a 42 × 60 pica trim size, the image area is 24 × 41 picas
with margins of 5p3 inside, 5p6 top, 6p9 outside, 7p6 bottom

Chaper titles are bold 18/21.6,
subheads are bold 12/14.4, italic;
body text is 11/14; extracts are 10.5/13 with 0p8 above and below;
running headers and footers are 9.5;
footnotes are 9/10.8 with 0p3 separation;
references are 10/12.5; the index is 8/8.8

Drops on the chapter title pages are
11p to the chapter title, and 18p6 to the text

231